Basics of Autodesk Inventor Nastran 2020 (Colored)

By
Gaurav Verma
Matt Weber
(CADCAMCAE Works)

ISBN # 978-1-988722-73-3

DEDICATION

To teachers, who make it possible to disseminate knowledge
to enlighten the young and curious minds
of our future generations

To students, who are the future of the world

THANKS

To my friends and colleagues

To my family for their love and support

Table of Contents

Preface viii
About Author x

Chapter 1 : Introduction to Simulation

Simulation **1-2**
Types of Analyses performed in Autodesk Inventor Nastran **1-2**
Static Analysis 1-2
Normal Modes Analysis 1-3
Buckling Analysis 1-4
Transient Response Analysis 1-6
Frequency Response Analysis 1-7
Impact Analysis 1-7
Random Response Analysis 1-7
Shock/Response Spectrum Analysis 1-8
Multi-Axial Fatigue Analysis 1-8
Vibrational Fatigue 1-8
Thermal analysis 1-8
Installing and Activating Inventor Nastran **1-9**
Performing Analysis (Overview) **1-10**
Applying Material 1-10
Applying Constraint 1-11
Applying Loads 1-12
Running Analysis 1-13

Chapter 2 : Basics of Analysis

Introduction **2-2**
Default Settings **2-2**
General Options 2-2
Setting Display Options 2-3
Nodes 2-3
Elements 2-4
Post-Processing 2-5
Report Generation 2-6
Starting A New Study **2-6**
Options tab 2-8
Editing Analysis **2-8**
Importing from Stress Analysis **2-9**
Applying Materials **2-9**
Creating New Material 2-10
Idealization **2-12**
Applying Connections between Assembly Components **2-13**
Applying Rod Connection 2-14
Applying Cable Connection 2-14
Applying Spring Connection 2-15
Applying Rigid Body Connection 2-17
Applying Bolt Connection 2-17

Offsetting Surface for Creating Shell Components **2-19**
Finding Thin Bodies **2-20**
Creating Midsurface **2-21**
Managing Structural Members **2-22**
Applying Constraints **2-23**
Applying Loads **2-24**
Applying Force 2-24
Applying Moment Load 2-26
Applying Distributed Load 2-27
Hydrostatic Load 2-27
Applying Pressure and Gravity 2-28
Applying Remote Force 2-28
Applying Bearing Load 2-30
Applying Rotational Force 2-31
Applying Enforced Motion Load 2-32
Applying Initial Condition 2-32
Applying Body Temperature 2-33
Applying Temperature 2-33
Applying Thermal Convection 2-34
Applying Thermal Radiation 2-36
Applying Heat Generation 2-36
Applying Heat Flux 2-37
Applying Loads based on Analysis Results 2-38
Defining Contact Sets **2-39**
Applying Automatic Contact Sets 2-39
Applying Manual Contacts 2-40
Setting Solver Options 2-43
Meshing **2-44**
Defining Mesh Settings 2-45
Mesh Table Parameters 2-46
Generating Mesh 2-47
Setting Mesh Control 2-47
Running Analysis **2-48**
Setting Plot Options **2-49**
Contour Options 2-50
Deform Options 2-50
Section View Options 2-50
Setting Part View Options 2-51
Vector Options 2-52
Animation Options 2-52
Visibility Options 2-53
Using Probe to Check Node Results 2-54

Chapter 3 : Static Analyses

Introduction **3-2**
Performing Linear Static Analysis on Part **3-2**
Applying Material 3-3
Applying Idealizations 3-4
Applying Constraints 3-4
Applying Force 3-4

Creating Mesh 3-5
Running Analysis 3-5
Performing Linear Static Analysis on an Assembly 3-6
Starting Linear Static Analysis 3-7
Applying Material 3-7
Applying Contacts 3-8
Applying Constraint 3-8
Applying Loads 3-8
Mesh Setting 3-9
Running Analysis 3-9
Performing Analysis on Frame Model 3-10
Performing Prestress Static Analysis 3-11
Non Linear Static Analysis 3-11
Theory Behind Non-Linear Analysis 3-12
Role of Time in Non-Linear Analysis 3-15
Performing Non Linear Static Analysis 3-15
Nonlinear Setup Parameters 3-16

Chapter 4 : Normal Modes Analysis

Introduction 4-2
Performing Normal Modes Analysis 4-3
Modal Setup Options 4-3
Performing Prestress Normal 4-5
Modes Analysis 4-5

Chapter 5 : Buckling Analysis

Introduction 5-2
Performing Linear Buckling Analysis 5-3
Performing Non-Linear Buckling Analysis 5-5

Chapter 6 : Transient Response Analysis

Introduction 6-2
Performing Direct Transient Response 6-3
Setting Damping Values 6-3
Dynamic Setup Parameters 6-4
Performing Modal Transient 6-5
Response Analysis 6-5
Performing Nonlinear Transient 6-7
Response Analysis 6-7

Chapter 7 : Impact Analysis

Introduction 7-2
Performing Impact Analysis 7-2
Defining Impact Parameters 7-4

Chapter 8 : Frequency Response Analysis

Introduction 8-2
Performing Direct Frequency Response Analysis 8-2
Performing Modal Frequency Response Analysis 8-4

Chapter 9 : Random Response Analysis

Introduction 9-2
Performing Random Response Analysis 9-2
Performing Shock/Response Spectrum Analysis 9-4

Chapter 10 : Fatigue Analyses

Introduction 10-2
Performing Multi-axial Fatigue Analysis 10-2
Fatigue Parameters 10-2
Performing Vibration Fatigue Analysis 10-4

Chapter 11 : Heat Transfer Analyses

Introduction 11-2
Performing Linear Steady State Heat Transfer Analysis 11-2
Performing Nonlinear Steady State Heat Transfer Analysis 11-4
Performing Nonlinear Transient Heat Transfer Analysis 11-5

Chapter 12 : Explicit Analyses

Introduction 12-2
Performing Explicit Dynamics Analysis 12-2
Applying Shell Idealization 12-2
Applying Impulse Load 12-3
Performing Explicit Quasi-Static Analysis 12-5

Chapter 13 : Fundamentals of FEA

Introduction 13-2
GENERAL DESCRIPTION OF THE METHODS 13-2
A BRIEF EXPLANATION OF FEA FOR A STRESS ANALYSIS PROBLEM 13-3
FINITE ELEMENT METHOD V/S CLASSICAL METHODS 13-5
FEM VS FINITE DIFFERENCE METHOD (FDM) 13-7
NEED FOR STUDYING FEM 13-7
Warning To Fea Package Users 13-8
Geometric Discontinuities 13-8
Discontinuity of Loads 13-9
Discontinuity of Boundary conditions 13-10
Material Discontinuity 13-10
REFINING MESH 13-11
Use Of Symmetry 13-11
HIGHER ORDER ELEMENTS V/S REFINED MESH 13-11

Practice Questions

Index I-1
Other Books by CADCAMCAE Works I-4

Preface

Autodesk Inventor® Nastran® software is a general purpose finite element analysis (FEA) tool embedded in Autodesk® Inventor. It is powered by the Autodesk® Nastran® solver and offers simulation capabilities that span across multiple analysis types, such as linear and nonlinear stress, dynamics, and heat transfer.

Autodesk Inventor® Nastran® is available as a network license and serves multiple CAD platforms to provide a consistent user experience and eliminate the need for multiple single-platform simulation technologies. It delivers high-end simulation technology in a CAD-embedded workflow so you can make great products.

The **Basics of Autodesk Inventor Nastran 2020**, is a book to help professionals as well as students in learning basics of Finite Element Analysis via Autodesk Inventor Nastran. The book follows a step by step methodology. This book explains the background work running behind your simulation analysis screen. The book starts with introduction to simulation and goes through all the analyses tools of Autodesk Inventor Nastran with practical examples of analysis. Chapter on manual FEA ensure the firm understanding of FEA concepts. Some of the salient features of this book are:

In-Depth explanation of concepts

Every new topic of this book starts with the explanation of the basic concepts. In this way, the user becomes capable of relating the things with real world.

Topics Covered

Every chapter starts with a list of topics being covered in that chapter. In this way, the user can easy find the topic of his/her interest easily.

Instruction through illustration

The instructions to perform any action are provided by maximum number of illustrations so that the user can perform the actions discussed in the book easily and effectively. There are about 300 illustrations that make the learning process effective.

Tutorial point of view

The book explains the concepts through the tutorial to make the understanding of users firm and long lasting. Each chapter of the book has tutorials that are real world projects.

Project

Free projects and exercises are provided to students for practicing.

For Faculty

If you are a faculty member, then you can ask for video tutorials on any of the topic, exercise, tutorial, or concept.

Formatting Conventions Used in the Text

All the key terms like name of button, tool, drop-down etc. are kept bold.

Free Resources

Link to the resources used in this book are provided to the users via email. To get the resources, mail us at ***cadcamcaeworks@gmail.com*** or ***info@cadcamcaeworks.com*** with your contact information. With your contact record with us, you will be provided latest updates and informations regarding various technologies. The format to write us e-mail for resources is as follows:

Subject of E-mail as ***Application for resources of ....Book.***
You can give your information below to get updates on the book.
Name:
Course pursuing/Profession:
Contact Address:
E-mail ID:

For Any query or suggestion

If you have any query or suggestion please let us know by mailing us on ***cadcamcaeworks@gmail.com*** or ***info@cadcamcaeworks.com***. Your valuable constructive suggestions will be incorporated in our books and your name will be addressed in special thanks area of our books.

About Author

Gaurav Verma is a Mechanical Design Engineer with deep knowledge of CAD, CAM and CAE field. He has an experience of more than 10 years on CAD/CAM/CAE packages. He has delivered presentations in Autodesk University Events on AutoCAD Electrical and Autodesk Inventor. He is an active member of Autodesk Knowledge Share Network. He has provided content for Autodesk Design Academy. He is also working as technical consultant for many Indian Government organizations for Skill Development sector. He has authored books on SolidWorks, Mastercam, Creo Parametric, Autodesk Inventor, Autodesk Fusion 360, and many other CAD-CAM-CAE packages. He has developed content for many modular skill courses like Automotive Service Technician, Welding Technician, Lathe Operator, CNC Operator, Telecom Tower Technician, TV Repair Technician, Casting Operator, Maintenance Technician and about 50 more courses. He has his books published in English, Russian and Hindi worldwide.

He has trained many students on mechanical, electrical and civil areas of CAD-CAM-CAE. He has trained students online as well as offline. He has a small workshop of 20 CNC and VMC machines where he tests his CAM skills on different Automotive components. He is providing consultant services to more than 15 companies worldwide. You can contact the author directly at cadcamcaeworks@gmail.com

This page is left blank intentionally

This page is left blank intentionally

Chapter 1

Introduction to Simulation

Topics Covered

The major topics covered in this chapter are:

- ***Simulation***
- ***Types of Analyses performed in Autodesk Inventor Nastran***
- ***FEA***
- ***Activating Autodesk Inventor Nastran***

SIMULATION

Simulation is the study of effects caused on an object due to real-world loading conditions. Computer Simulation is a type of simulation which uses CAD models to represent real objects and it applies various load conditions on the model to study the real-world effects. Autodesk Inventor Nastran is one of the Computer Simulation programs available in the market. In Autodesk Inventor Nastran, we apply loads on a constrained model under predefined environmental conditions and check the result(visually and/or in the form of tabular data). The types of analyses that can be performed in Autodesk Inventor Nastran are given next.

TYPES OF ANALYSES PERFORMED IN AUTODESK INVENTOR NASTRAN

Autodesk Inventor Nastran performs almost all the analyses that are generally performed in Industries. These analyses and their uses are given next.

Static Analysis

This is the most common type of analysis we perform. In this analysis, loads are applied to a body due to which the body deforms and the effects of the loads are transmitted throughout the body. To absorb the effect of loads, the body generates internal forces and reactions at the supports to balance the applied external loads. These internal forces and reactions cause stress and strain in the body. Static analysis refers to the calculation of displacements, strains, and stresses under the effect of external loads, based on some assumptions. The assumptions are as follows.

- All loads are applied slowly and gradually until they reach their full magnitudes. After reaching their full magnitudes, load will remain constant (i.e. load will not vary against time).
- Linearity assumption: The relationship between loads and resulting responses is linear. For example, if you double the magnitude of loads, the response of the model (displacements, strains and stresses) will also double. You can make linearity assumption if:

1. All materials in the model comply with Hooke's Law that is stress is directly proportional to strain.
2. The induced displacements are small enough to ignore the change in stiffness caused by loading.
3. Boundary conditions do not vary during the application of loads. Loads must be constant in magnitude, direction, and distribution. They should not change while the model is deforming.

Linear Static Analysis

If the above assumptions are valid for your analysis, then you can perform **Linear Static Analysis**. For example, a cantilever beam fixed at one end and force applied on other end; refer to Figure-1.

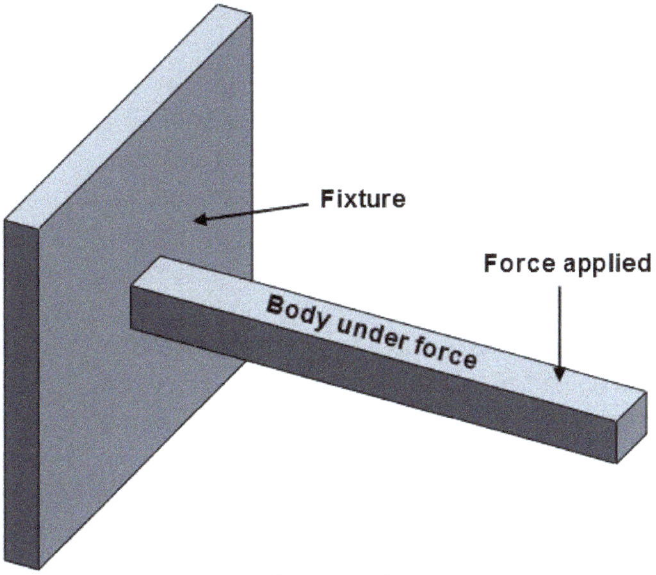

Figure-1. Linear static analysis example

Nonlinear Static Analysis

If the above assumptions are not valid, then you need to perform the **Non-Linear Static analysis**. For example, an object attached with a spring being applied under forces; refer to Figure-2.

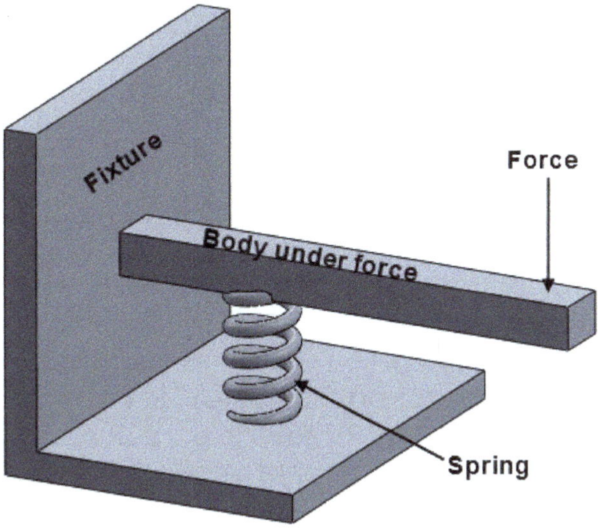

Figure-2. Non-linear static analysis example

Prestress Static Analysis

The Prestress static analysis is performed when you have model already prestressed and want to apply additional loads.

Normal Modes Analysis

The Normal Modes Analysis also called harmonic analysis is used to find natural frequencies. By its very nature, vibration involves repetitive motion. Each occurrence of a complete motion sequence is called a "cycle." Frequency is defined as so many cycles in a given time period. "Cycles per seconds" or "Hertz". Individual parts have what engineers call "natural" frequencies. For example, a violin string at a certain tension will vibrate only at a set number of frequencies, which is why you can produce specific musical tones. There is a base frequency at which the entire string is going back and forth in a simple bow shape.

Harmonics and overtones occur because individual sections of the string can vibrate independently within the larger vibration to form different shapes. These various shapes are called “modes”. The base frequency is said to vibrate in the first mode, and so on up the ladder. Each mode shape will have an associated frequency. Higher mode shapes have higher frequencies. The most disastrous kinds of consequences occur when a power-driven device such as a motor for example, produces a frequency at which an attached structure naturally vibrates. This event is called “resonance.” If sufficient power is applied, the attached structure will be destroyed. Note that ancient armies, which normally marched “in step,” were taken out of step when crossing bridges. Should the beat of the marching feet align with a natural frequency of the bridge, it could fall down. Engineers must design in such a way that resonance does not occur during regular operation of machines. This is a major purpose of Normal Modes Analysis. Ideally, the first mode has a frequency higher than any potential driving frequency. Frequently, resonance cannot be avoided, especially for short periods of time. For example, when a motor comes up to speed it produces a variety of frequencies. So it may pass through a resonant frequency.

Buckling Analysis

The Buckling Analysis is performed to check sudden failure of structure. If you press down on an empty soft drink can with your hand, not much will seem to happen. If you put the can on the floor and gradually increase the force by stepping down on it with your foot, at some point it will suddenly squash. This sudden scrunching is known as “buckling.”

Models with thin parts tend to buckle under axial loading. Buckling can be defined as the sudden deformation, which occurs when the stored membrane(axial) energy is converted into bending energy with no change in the externally applied loads. Mathematically, when buckling occurs, the total stiffness matrix becomes singular.

In the normal use of most products, buckling can be catastrophic if it occurs. The failure is not one because of stress but geometric stability. Once the geometry of the part starts to deform, it can no longer support even a fraction of the force initially applied. The worst part about buckling for engineers is that buckling usually occurs at relatively low stress values for what the material can withstand. So they have to make a separate check to see if a product or part thereof is okay with respect to buckling.

Slender structures and structures with slender parts loaded in the axial direction buckle under relatively small axial loads. Such structures may fail in buckling while their stresses are far below critical levels. For such structures, the buckling load becomes a critical design factor. Stocky structures, on the other hand, require large loads to buckle, therefore buckling analysis is usually not required.

Buckling almost always involves compression; refer to Figure-3. In mechanical engineering, designs involving thin parts in flexible structures like airplanes and automobiles are susceptible to buckling. Even though stress can be very low, buckling of local areas can cause the whole structure to collapse by a rapid series of 'propagating buckling'. Buckling analysis calculates the smallest (critical) loading required buckling a model. Buckling loads are associated with buckling modes. Designers are usually interested in the lowest mode because it is associated with the lowest critical load. When buckling is the critical design factor, calculating multiple buckling modes helps in locating the weak areas of the model. This may prevent the occurrence of lower buckling modes by simple modifications.

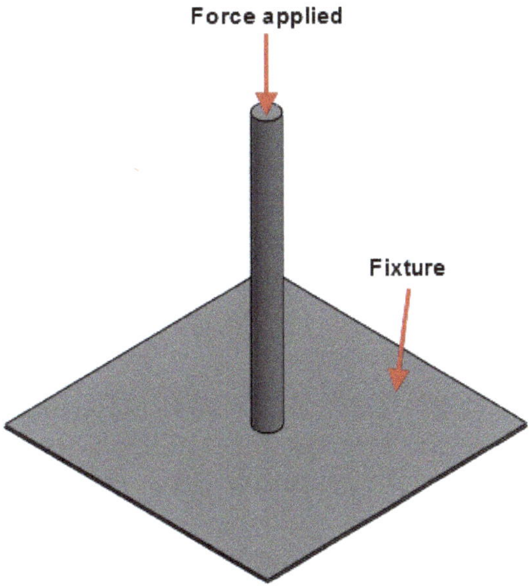

Figure-3. Buckling example

Linear Buckling Analysis

Linear-buckling analysis (also called eigenvalue-based buckling analysis) is in many ways similar to modal analysis. Linear buckling is the most common type of analysis and is easy to execute, but it is limited in the results it can provide.

Linear-buckling analysis calculates buckling load magnitudes that cause buckling and associated buckling modes. FEA programs provide calculations of a large number of buckling modes and the associated buckling-load factors (BLF). The BLF is expressed by a number which the applied load must be multiplied by (or divided — depending on the particular FEA package) to obtain the buckling-load magnitude.

The buckling mode presents the shape the structure will assume when it buckles in a particular mode, but says nothing about the numerical values of the displacements or stresses. The numerical values can be displayed, but are merely relative. This is in close analogy to modal analysis, which calculates the natural frequency and provides qualitative information on the modes of vibration (modal shapes), but not on the actual magnitude of displacements.

Nonlinear Buckling Analysis

The nonlinear-buckling analysis requires a load to be applied gradually in multiple steps rather than in one step as in a linear analysis. Each load increment changes the structure's shape, and this, in turn, changes the structure's stiffness. Therefore, the structure stiffness must be updated at each increment. In this approach, which is called the load control method, load steps are defined either by the user or automatically so the difference in displacement between the two consecutive steps is not too large.

Although the load-control method is used in most types of nonlinear analyses, it would be difficult to implement in a buckling analysis. When buckling happens, the structure undergoes a momentary loss of stiffness and the load control method would result in numerical instabilities. Nonlinear buckling analysis requires another way of controlling load application — the arc length control method. Here, points corresponding to consecutive load increments are evenly spaced along the load-displacement curve, which itself is constructed during load application.

In contrast to linear-buckling analysis, which only calculates the potential buckling shape with no quantitative values of importance, nonlinear analysis calculates actual displacements and stresses. To better understand the inner workings of nonlinear-buckling analysis, first consider what happens in running a nonlinear-buckling analysis on an idealized structure. Imagine a perfectly round and perfectly straight column under a perfectly aligned compressive load. Theoretically, buckling will never happen, but in actuality, buckling will take place because of imperfections in the geometry, loads, and supports.

Transient Response Analysis

Transient response analysis is the most general method for computing forced dynamic response. The purpose of a transient response analysis is to determine the behavior of a structure subjected to time-varying excitation. The transient excitation is explicitly defined in the time domain. The loads applied to the structure are known at each instant in time. Loads can be in the form of applied forces and enforced motions. The results obtained from a transient response analysis are typically displacements, velocities, and accelerations of grid points, and forces and stresses in elements, at each output time step. Depending upon the structure and the nature of the loading, two different numerical methods can be used for a transient response analysis:

Direct Transient Response Analysis

Direct Transient Response Analysis calculates the response of a system to a load over time. The load applied to the system can vary over time or simply be an initial condition that is allowed to evolve over time. This method may be more efficient for models where high-frequency excitation require the extraction of a large number of modes. Also, if structural damping is used, the direct method should be used.

Modal Transient Response Analysis

Modal Transient Response Analysis is an alternate technique available for dynamics that utilizes the mode shapes of the structure, reduces the solution degrees of freedom, and can significantly impact the run time. This approach replaces the physical degrees of freedom with a reduced number of modal degrees of freedom. Fewer degrees of freedom mean a faster solution. This can be a big time saver for transient models with a large number of time steps. Because modal transient response analysis uses the mode shapes of a structure, this analysis is a natural extension of normal modes analysis.

Nonlinear Transient Response Analysis

A nonlinear transient analysis requires both dynamic and nonlinear setup steps. Autodesk Inventor Nastran solves both analyses essentially simultaneously, making it one of the most complex yet exciting solution types in FEA.

An important element to having a stable nonlinear transient (NLT) solution is to provide damping in the model. There are two types of damping that can be applied in NLT solutions:

Global damping value: is specified using a PARAM,G followed by a PARAM,W3 which defines the frequency at which to apply the damping.

Material based damping is defined on each material card directly. PARAM,W4 is needed to define the frequency at which to apply the material based damping. Note that the units of W3 and W4 are radians per unit time.

Frequency Response Analysis

The frequency response analysis is used to compute structural response of model to steady-state oscillatory excitation. In frequency response analysis, the excitation is explicitly defined in the frequency domain. Excitations can be in the form of applied forces and enforced motions (displacements, velocities, or accelerations). There are two types of frequency response analysis:

Direct Frequency Response Analysis

In Direct Frequency Response Analysis the structural response is computed at discrete excitation frequencies by solving a set of coupled matrix equations using complex algebra. The direct method may be more efficient for models where high-frequency excitation require the extraction of a large number of modes.

Modal Frequency Response Analysis

Modal Frequency Response Analysis is an alternate method to compute frequency response. This method uses the mode shapes of the structure to uncouple the equations of motion (when no damping or only modal damping is used) and, depending on the number of modes computed and retained, reduce the problem size. Both of these factors tend to make modal frequency response analysis computationally more efficient than direct frequency response analysis. It is used for large models where a large number of solution frequencies are specified. This method replaces the physical degrees of freedom (DOF) with a reduced number of modal degrees of freedom. Fewer degrees of freedom mean a faster solution. Because modal frequency response analysis uses the mode shapes of a structure, modal frequency response analysis is a natural extension of normal modes analysis.

Impact Analysis

The Impact analysis is used to perform drop-test and projectile impact studies. This study simulate the effect of dropping a part or an assembly on a rigid or flexible floor. To perform this study, the floor is considered as planar and flat. The forces that are considered automatically for this study are gravity and impact reaction.

Random Response Analysis

Engineers use this type of analysis to find out how a device or structure responds to steady shaking of the kind you would feel riding in a truck, rail car, rocket (when the motor is on), and so on. Also, things that are riding in the vehicle, such as on-board electronics or cargo of any kind, may need Random Vibration Analysis. The vibration generated in vehicles from the motors, road conditions, etc. is a combination of a great many frequencies from a variety of sources and has a certain "random" nature. Random Vibration Analysis is used by mechanical engineers who design various kinds of transportation equipment.

Shock/Response Spectrum Analysis

Engineers use this type of analysis to find out how a device or structure responds to sudden forces or shocks. It is assumed that these shocks or forces occur at boundary points, which are normally fixed. An example would be a building, dam or nuclear reactor when an earthquake strikes. During an earthquake, violent shaking occurs. This shaking transmits into the structure or device at the points where they are attached to the ground (boundary points).

Mechanical engineers who design components for nuclear power plants must use response spectrum analysis as well. Such components might include nuclear reactor parts, pumps, valves, piping, condensers, etc. When an engineer uses response spectrum analysis, he is looking for the maximum stresses or acceleration, velocity and displacements that occur after the shock. These in turn lead to maximum stresses.

Multi-Axial Fatigue Analysis

The Multi-Axial Fatigue Analysis is used to check the effect of continuous loading and unloading of forces on a body. The base element for performing fatigue analysis are results of static, nonlinear, or time history linear dynamic studies.

Vibrational Fatigue

The Vibrational Fatigue Analysis is used to check the effect of vibrational loading and unloading on the body.

Thermal analysis

There are three mechanisms of heat transfer. These mechanisms are Conduction, Convection, and Radiation. Thermal analysis calculates the temperature distribution in a body due to some or all of these mechanisms. In all three mechanisms, heat flows from a higher-temperature medium to a lower temperature one. Heat transfer by conduction and convection requires the presence of an intervening medium while heat transfer by radiation does not.

There are two modes of heat transfer analysis.

Steady State Thermal Analysis

In this type of analysis, we are only interested in the thermal conditions of the body when it reaches thermal equilibrium, but we are not interested in the time it takes to reach this status. The temperature of each point in the model will remain unchanged until a change occurs in the system. At equilibrium, the thermal energy entering the system is equal to the thermal energy leaving it. Generally, the only material property that is needed for steady state analysis is the thermal conductivity.

Transient Thermal Analysis

In this type of analysis, we are interested in knowing the thermal status of the model at different instances of time. A thermos designer, for example, knows that the temperature of the fluid inside will eventually be equal to the room temperature(steady state), but he is interested in finding out the temperature of the fluid as a function of time. In addition to the thermal conductivity, we also need to specify density, specific heat, initial temperature profile, and the period of time for which solutions are desired.

Till this point, we have learned the basic types of analyses that can be performed in Inventor Nastran. Now, we will discuss the procedure of activating Inventor Nastran and starting a basic analysis.

INSTALLING AND ACTIVATING INVENTOR NASTRAN

The procedure to install and activate Inventor Nastran tool in Autodesk Inventor is given next.

- Open the link **https://www.autodesk.com/education/free-software/autodesk-inventor-nastran** in your default web browser. The web page will be displayed as shown in Figure-4.

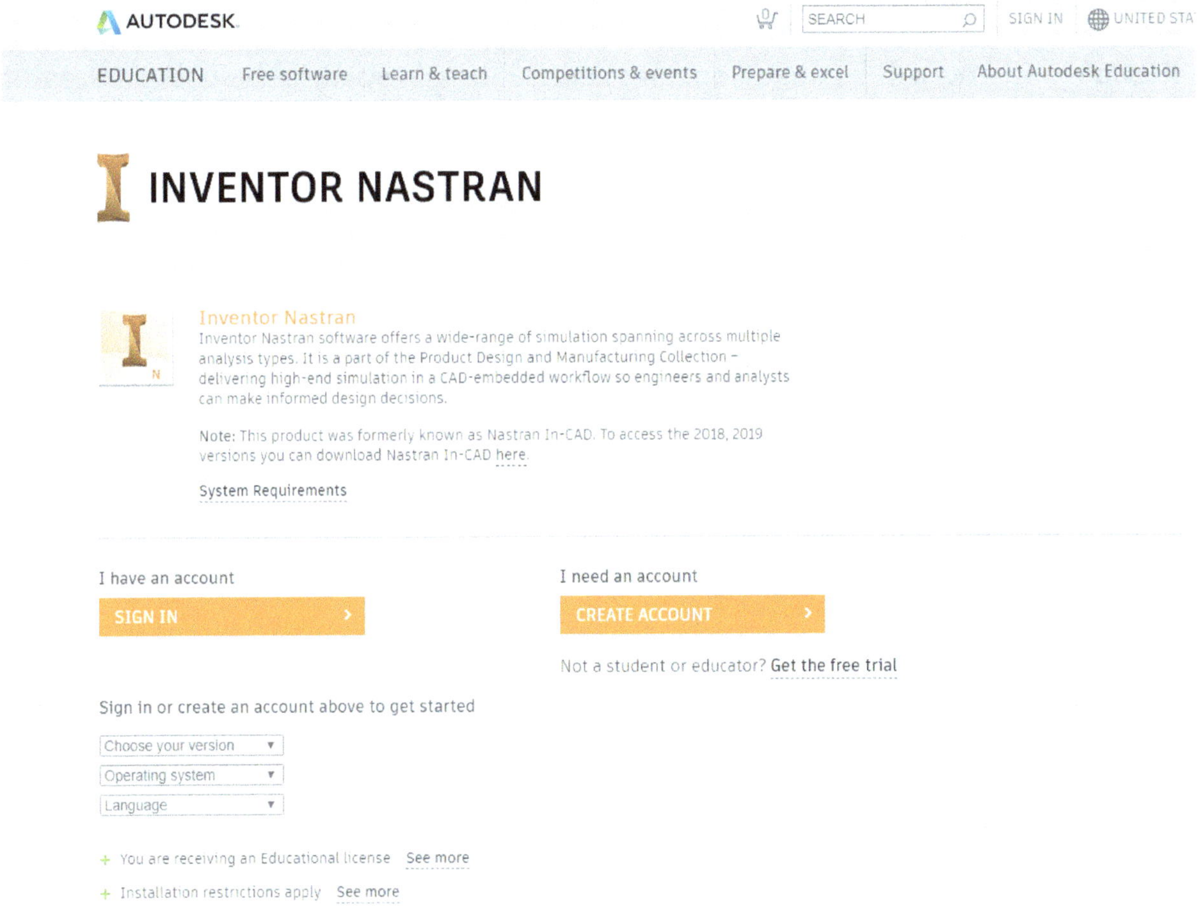

Figure-4. Inventor Nastran webpage

- Click on the **SIGN IN** button if you have an autodesk student/educator account already created otherwise click on the **CREATE ACCOUNT** button and create an educator/student account and then sign in.
- Select the desired version, operating system, and language in respective edit boxes; refer to Figure-5. Scroll down in the web page and click on the **INSTALL NOW** button. Install the software by following the instructions displayed.

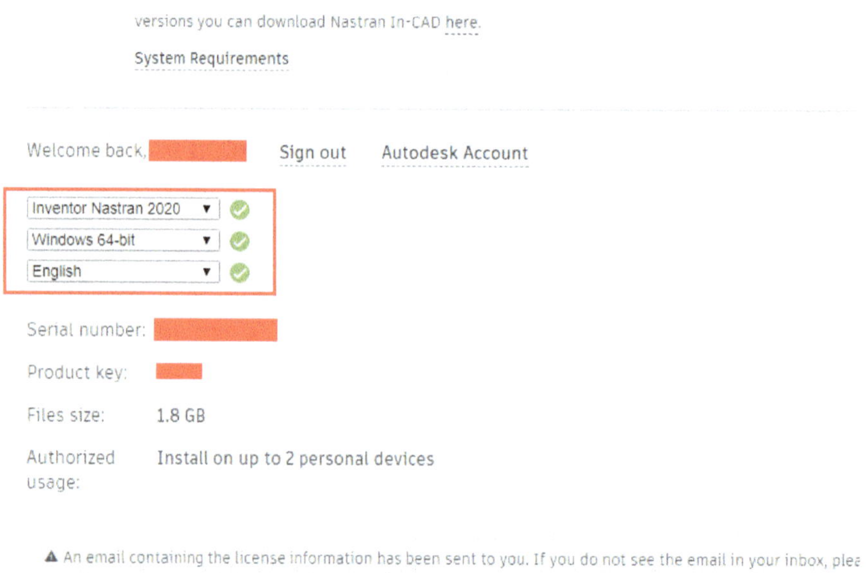

Figure-5. Selecting software options in webpage

- Start Autodesk Inventor and open the model on which you want to perform analysis. Click on the **Autodesk Inventor Nastran** tool from the **Begin** panel in the **Environments** tab of **Ribbon**. The **Autodesk Nastran** tab will be added in the **Ribbon** and you will be asked specify license detail. Specify the license detail for the software. The refer to Figure-6.

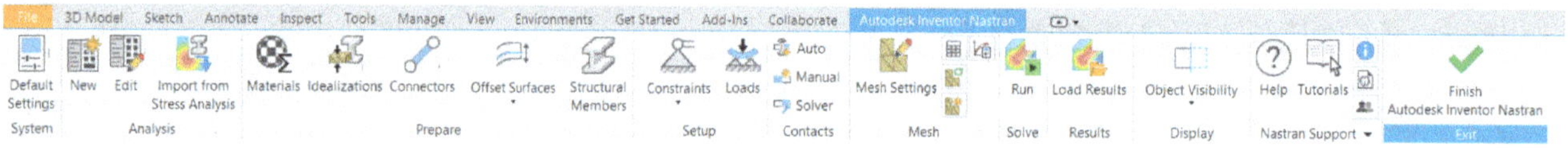

Figure-6. Autodesk Nastran tab

PERFORMING ANALYSIS (OVERVIEW)

There are various steps involved in performing analysis like applying material, assigning constraints, applying loads, applying mesh settings, and performing analysis. These steps are given next.

Applying Material

Applying material is an important step in analysis. Material properties can very much affect the result of analysis. The procedure to apply material is given next.

- Click on the **Materials** tool from the **Prepare** panel in the **Autodesk Nastran** tab of **Ribbon**. The **Material** dialog box will be displayed; refer to Figure-7.

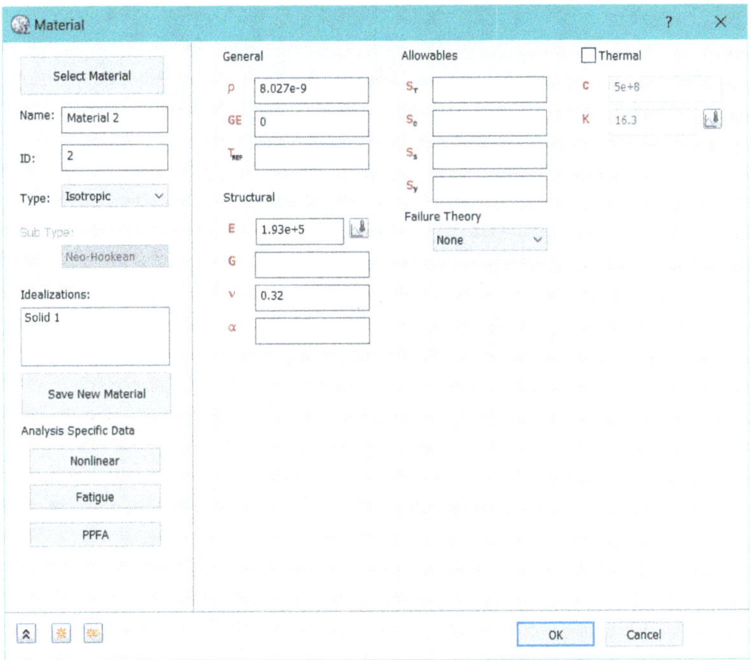

Figure-7. Material dialog box

- Set the desired parameters in the dialog box for material properties.
- If you want to select the material from library then click on the **Select Material** button. The **Material DB** dialog box will be displayed; refer to Figure-8.

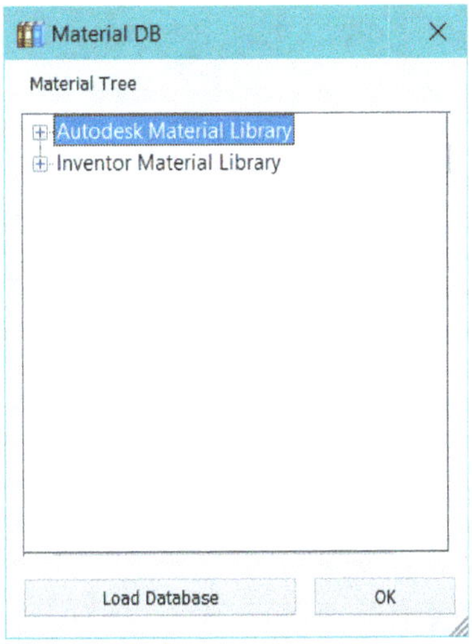

Figure-8. Material DB dialog box

- Expand the node and select the desired material. After selecting material, click on the **OK** button. The **Material** dialog box will be displayed again.
- Click on the **OK** button to apply material.

Applying Constraint

The constraints are used to restrict the motion of part during the analysis. The procedure to apply constraint is given next.

- Click on the **Constraints** tool from the **Setup** panel in the **Autodesk Nastran** tab of the **Ribbon**. The **Constraint** dialog box will be displayed; refer to Figure-9.

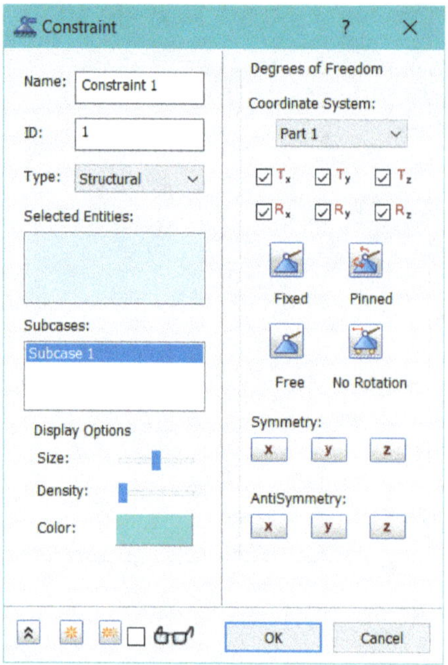

Figure-9. Constraint dialog box

- Select the desired option from the **Type** drop-down. There are three options in this drop-down; **Structural**, **Thermal**, and **Response Spectrum**. Select the **Structural** option if you want to apply structural constraints like restricting translation and rotation of selected object. Select the **Thermal** option if you want to apply temperature to selected face. Select the **Response Spectrum** option if you want to restrict moment or reaction force for selected objects.
- Click in the **Selected Entities** selection box and select the faces/edges/vertices that you want to constrained.
- Select the desired option from the **Degrees of Freedom** area.
- Set the desired display options from the **Display Options** area of the dialog box.
- After setting the desired parameters, click on the **OK** button.

Applying Loads

Loads are applied on the model to represent real load conditions for analysis. The procedure to apply loads is given next.

- Click on the **Loads** tool from the **Setup** panel in the **Autodesk Nastran** tab. The **Load** dialog box will be displayed; refer to Figure-10.

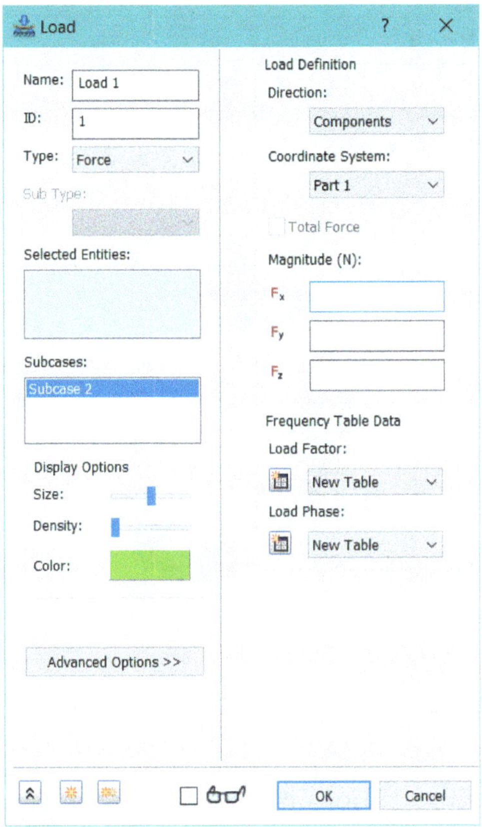

Figure-10. Load dialog box

- Select the desired option from the **Type** drop-down. There are various options like force, moment, pressure, distributed load etc. to define different types of loads.
- Select the **Force** option from the drop-down and set the desired value of force in the **Magnitude (N)** edit box(es).
- Click in the **Selected Entities** selection box and select the faces/edges/vertices to apply load.
- Click on the **OK** button to apply load.

Running Analysis

- Click on the **Generate Mesh** tool from the **Mesh** panel in the **Autodesk Inventor Nastran** tab of **Ribbon**. The preview of mesh will be displayed; refer to Figure-11.

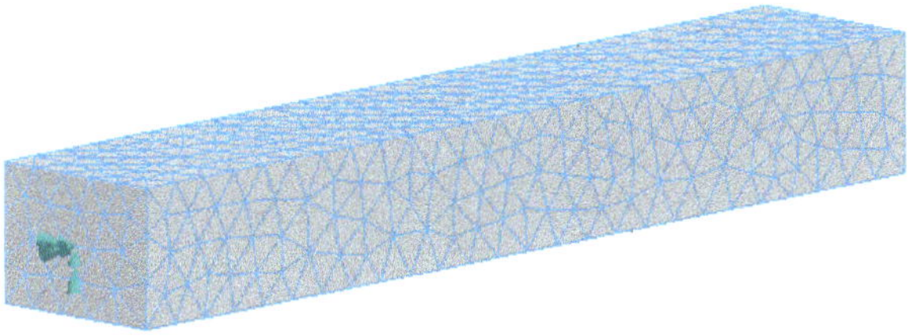

Figure-11. Meshing created

- Click on the **Run** tool from the **Solve** panel in **Autodesk Inventor Nastran** tab of the **Ribbon**. If you have not saved the file then **Save As** dialog box will be displayed. Save the file at desired location. Once the analysis is complete, the results of analysis will be displayed; a message box will be displayed telling you that analysis solution has been generated. Click on the **OK** button from the dialog box; refer to Figure-12.

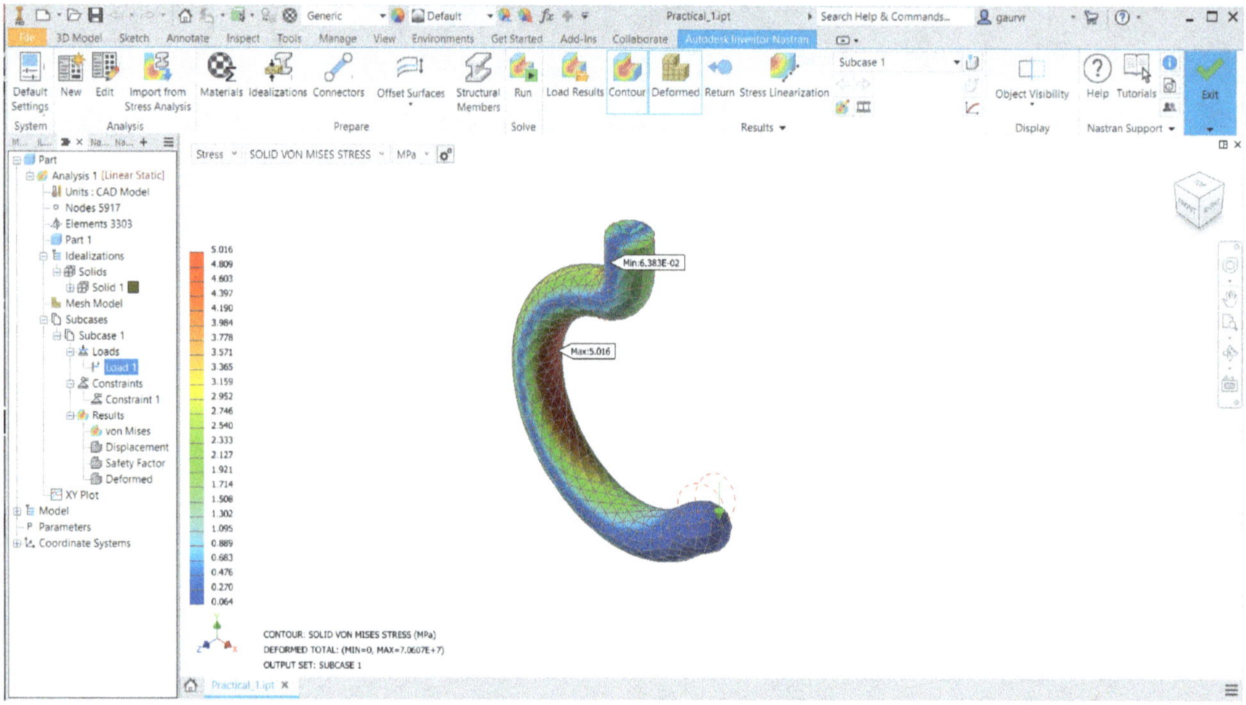

Figure-12. Result of analysis

You will learn more about these tools in the next chapter.

Chapter 2

Basics of Analysis

Topics Covered

The major topics covered in this chapter are:

- ***Introduction***
- ***Starting Analysis***
- ***Default Settings***
- ***Starting New Study***
- ***Editing Analysis Parameters***
- ***Importing Data from Stress Analysis***
- ***Applying and Editing Material***
- ***Idealization***
- ***Applying Connections to Components in Assemblies***
- ***Generating Surfaces for Shell Components***
- ***Managing Structural Members***
- ***Applying Loads and Constraints***
- ***Defining Contacts***
- ***Generating Meshes***
- ***Running Analyses and Generating Plots***

INTRODUCTION

In previous chapter, you learned the basic process of performing analysis in Autodesk Inventor Nastran. Now, we will discuss each of the tool in **Autodesk Inventor Nastran** tab one by one.

DEFAULT SETTINGS

The **Default Settings** tool is used to set display options and other basic settings of the dialog box. The procedure to use this tool is given next.

- Click on the **Default Settings** tool from the **System** panel in the **Autodesk Inventor Nastran** tab of the **Ribbon**. The **Default Settings** dialog box will be displayed; refer to Figure-1.

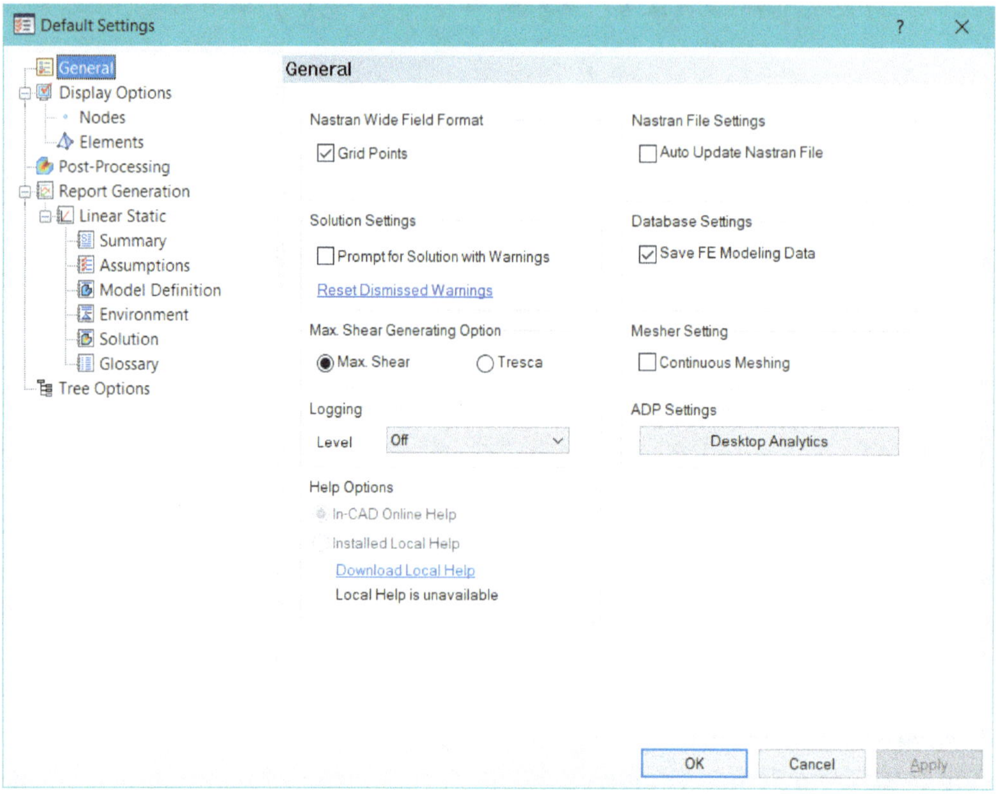

Figure-1. Default Settings dialog box

General Options

- Click on the **General** option from the left area to set default display options. The options will be displayed as shown in above figure.
- Select the **Grid Points** check box to display grid points of the model.
- Select the **Prompt for Solution with Warnings** check box to let system prompt for solutions if there are any warnings.
- Select the **Max Shear** radio button if you want find maximum shear stress of the model in result. Select the **Tresca** radio button if you want to find out shell/solid tresca results.
- Select the **Auto Update Nastran File** check box to automatically update the nastran files based on specified parameters.
- Select the **Save FE Modeling Data** check box to save finite element data separately in organized structure.

- Select the **Continuous Meshing** check box to mesh shell objects automatically during editing.
- Select the desired option from the **Level** drop-down in the **Logging** area. Select **Crash** option if you want the system to generate log of software events when the software crashes. Select the **Verbose** option if you want the system to generate log of all events of the software.
- Select the **Installed Local Help** radio button from **Help Options** area of dialog box if you have installed local copy of help and want to use it for reference.
- Click on the **Desktop Analytics** button from the dialog box if you want to enable or disable data collection by Autodesk. On clicking this button, the **Data collection and use** dialog box will be displayed. Select the check box in the dialog box if you want to enable data collection and click on the **OK** button.

Setting Display Options

- Click on the **Display Options** node from the left area of the dialog box. The options in the dialog box will be displayed as shown in Figure-2.

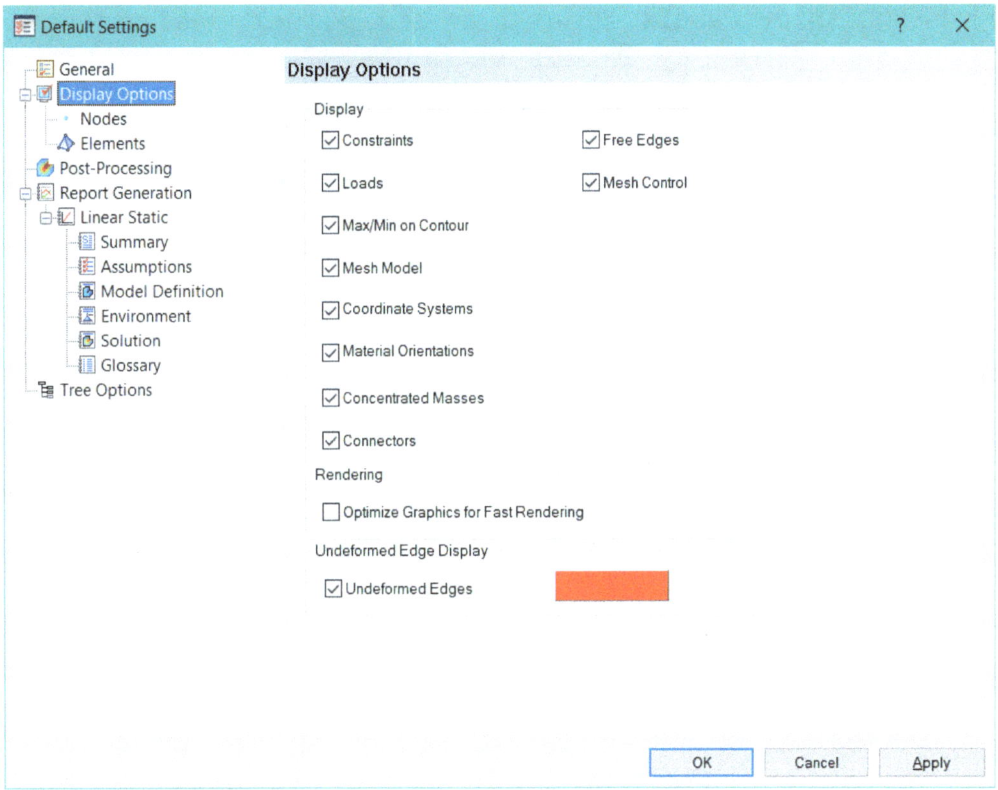

Figure-2. Display Options page

- Select the desired check boxes from the **Display Options** area of the dialog box to set which entities of model to be displayed.

Nodes

- Select the **Nodes** option from the left area of the dialog box. The dialog box will be displayed as shown in Figure-3.

- Select the **Display Nodes** check box to display nodes in the model.

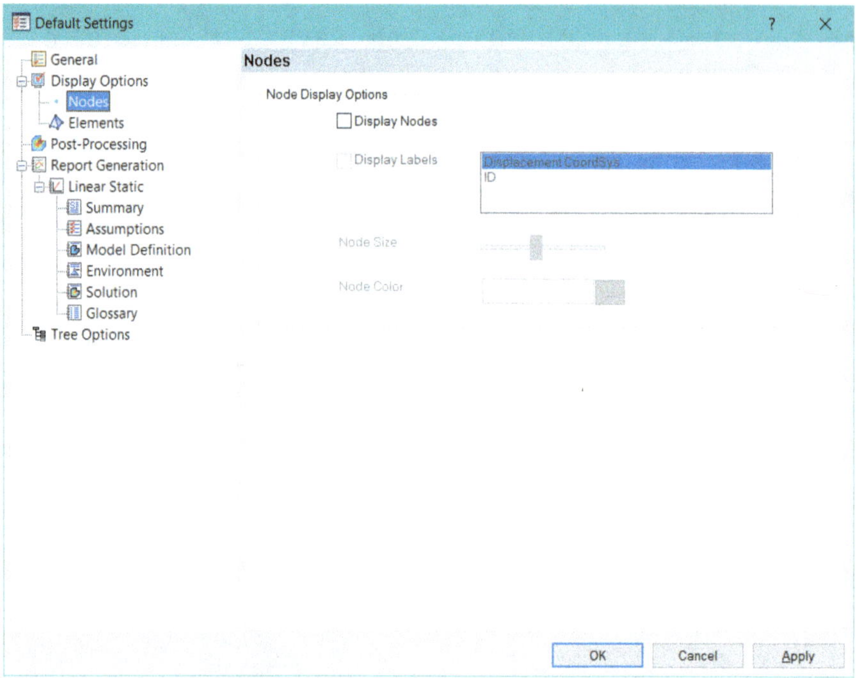

Figure-3. Nodes display options

- Select the **Display Labels** check box and select desired option from the list box to display labels on nodes.
- Set the desired parameters for node size and color from this page.

Elements

- Select the **Elements** option from the **Default Settings** dialog box to set the options for display of element labels. The dialog box will be displayed as shown in Figure-4.
- Select the **Display Labels** check box and select the label types to be displayed.

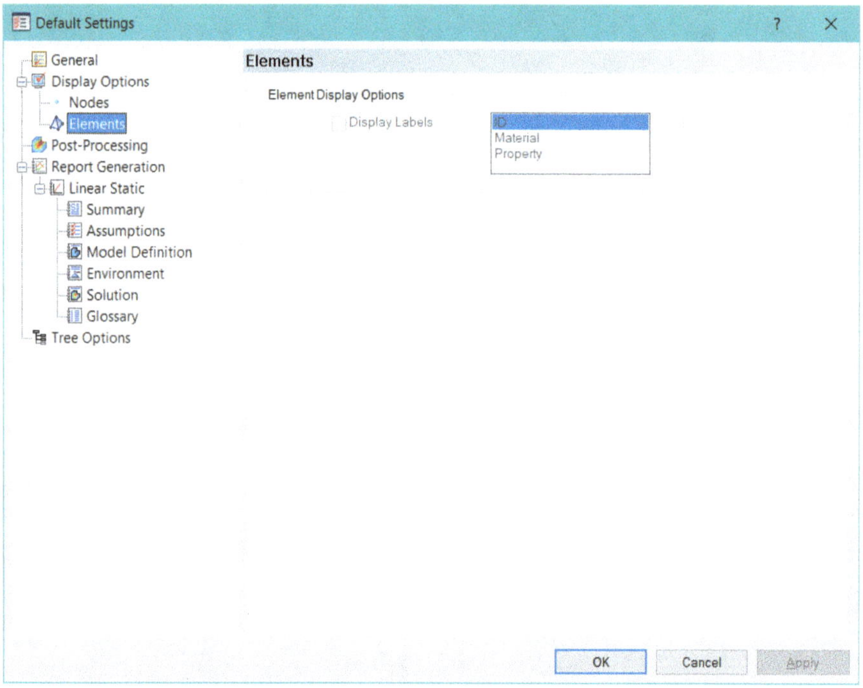

Figure-4. Elements display options

Post-Processing

- Select the **Post-Processing** option from the left area of the dialog box. The options in the dialog box will be displayed as shown in Figure-5.

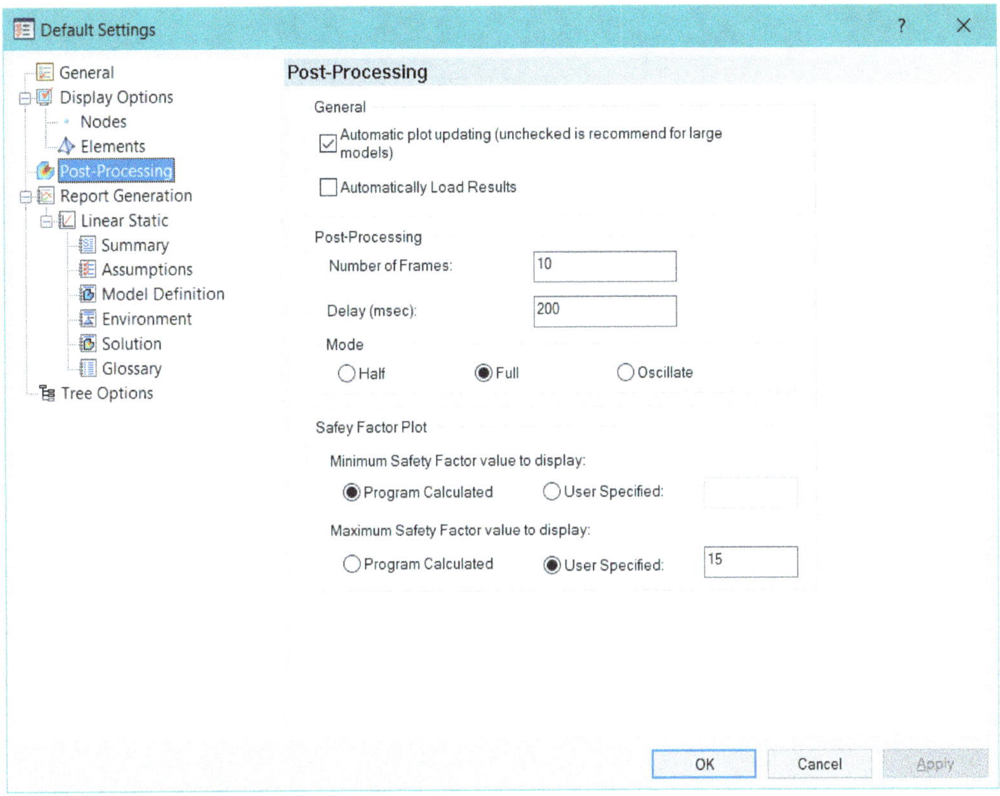

Figure-5. Post-Processing display options

- Select the **Automatic plot updating** check box to automatically update all plots after performing analysis.
- Select the **Automatically Load Results** check box to automatically load the results after performing analysis.
- Specify the number of frames to be created per second for animation in the **Number of Frames** edit box.
- Specify the desired value of microseconds delay between frames of animation in the **Delay (msec)** edit box.
- Select the desired option from the **Mode** area. If you select the **Half** option then the animation will start at unloaded/undeformed state, gradually advance to the fully loaded/deformed results, and then stop. If you select the **Full** option then animation will start at the unloaded/undeformed state, gradually advancing to the fully loaded/deformed results, and then reversing (gradually returning to the unloaded/undeformed state). If you **Oscillate** option then first half of the animation is identical to the **Full** option. After that, a second full-cycle is added using a mirror image of the calculated deformed shape. This option is useful for visualizing vibration modes. For example, if the top of a tower bends to the right when determining the mode shape, it will also bend to the left for the second half of a full oscillation cycle.
- Set the desired option for minimum and maximum safety factor value in the **Safety Factor Plot** area of the dialog box.

Report Generation

- Click on the **Report Generation** option from the left area of the dialog box. The options in the dialog box will be displayed as shown in Figure-6.

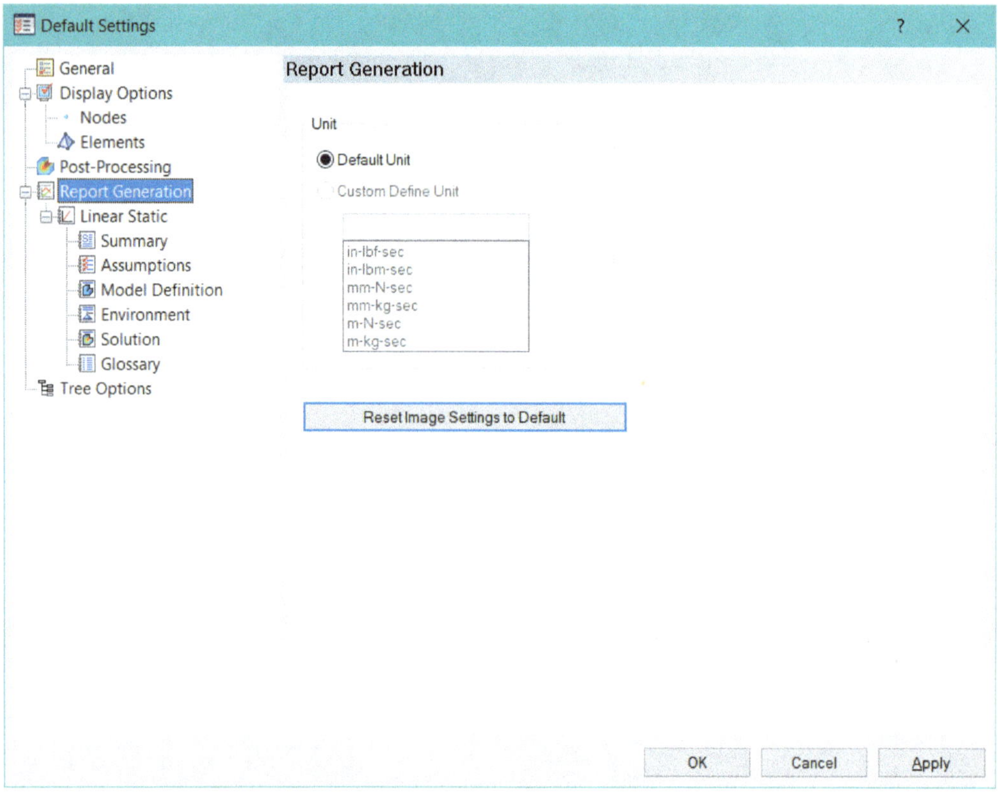

Figure-6. Report Generation options

- Select the desired unit system from the dialog box.
- Set the options for current analysis as required by using the nodes from the left box.
- After setting desired parameters, click on the **OK** button.

STARTING A NEW STUDY

The **New** tool in the **Analysis** panel of **Autodesk Inventor Nastran** tab is used to create new simulation study. The procedure to use this tool is given next.

- Click on the **New** tool from the **Analysis** panel in the **Autodesk Inventor Nastran** tab of the **Ribbon**. The **Analysis** dialog box will be displayed; refer to Figure-7.

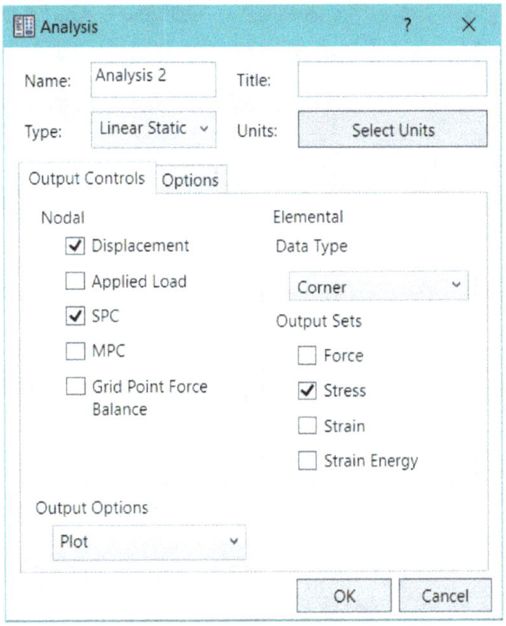

Figure-7. Analysis dialog box

- Specify the desired name of analysis in **Name** edit box and specify the desired title for documentation of current analysis in the **Title** edit box.
- Select the desired type of analysis from the **Type** drop-down like Linear Static, Normal Modes, and so on.
- Click on the **Select Units** button to set units for the analysis. The **Units** dialog box will be displayed; refer to Figure-8. Select the desired option from the **Unit System** drop-down and click on the **OK** button. The selected unit will be set and the **Analysis** dialog box will be displayed again.

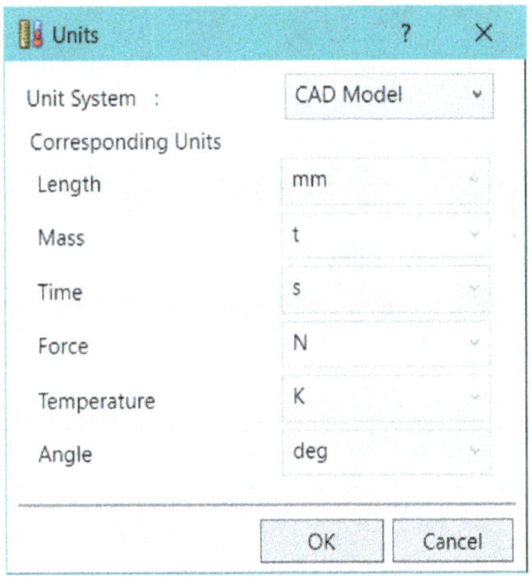

Figure-8. Units dialog box

- Select the desired check box(es) from the **Nodal** area of the dialog box to set as output control parameter. The selected parameter will be plotted in output files. For example, select the **Displacement** check box to use displacement as controlling parameter for analysis. So, apart from general data model with grid points and elements for analysis, there will also be a matrix of displacement for solving the equations. You can find the equations in .nas file generated after solving analysis in the same folder where your model file is stored.

- Select the desired option from the **Output Options** drop-down to define how output of analysis will be generated. Select the **Plot** option to generate nastran plot file (.NAS file) for result. Select the **Print** option if you want to generate additional result output file(.OUT file). Select the **Punch** option if you want to generate a punch file (.PCH) for results. Select the **Punch and Plot** option if you want to generate both punch(.PCH) and nastran output (.NAS) files for results.
- Select the desired option from the **Data Type** drop-down to define what elemental data will be generated. Select the **Corner** option to use results at the nodes of elements for calculating final result. Select the **Centroidal** option from the drop-down to use results calculated at centroids of elements for calculating final result.
- From the **Output Sets** area, select the desired check boxes for parameters you want to be generated in result files.

Options tab

- Click on the **Options** tab in the dialog box. The options will be displayed as shown in Figure-9.

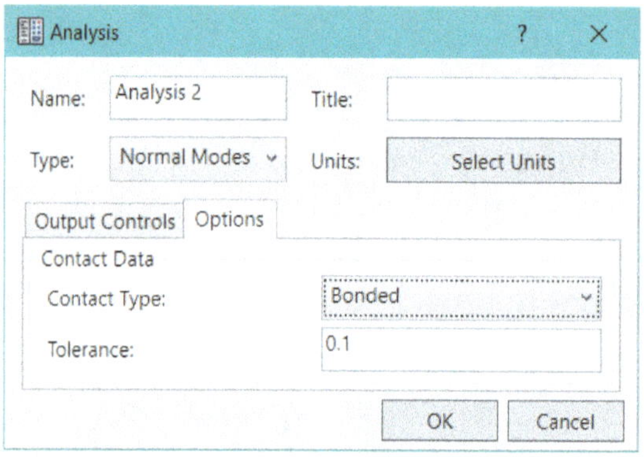

Figure-9. Options tab

- Select the desired option from the **Contact Type** drop-down to define the way two or more bodies will be joined for the analysis. The option set here will be default for the analysis. Note that you can later modify contacts between various components.
- Set the desired tolerance value in the **Tolerance** edit box. If two components have separation distance within the specified tolerance range then the selected contact type will be applied automatically.
- Click on the **OK** button from the dialog box to apply parameters.

EDITING ANALYSIS

The **Edit** tool in the **Analysis** panel is used to edit general parameters specified for the analysis. The procedure to use this tool is given next.

- Click on the **Edit** tool from the **Analysis** panel in the **Autodesk Inventor Nastran** tab of **Ribbon**. The **Analysis** dialog box will be displayed. The options of this dialog box have already been discussed.

IMPORTING FROM STRESS ANALYSIS

The **Import from Stress Analysis** tool is used to import the study parameters from Autodesk Inventor Stress Analysis. The procedure to use this tool is given next.

- Open a part in Autodesk Inventor on which you have earlier performed stress analysis.
- Start Autodesk Inventor Nastran environment as discussed earlier.
- Click on the **Import from Stress Analysis** tool from the **Analysis** panel in the **Autodesk Inventor Nastran** tab of the **Ribbon**. The analysis will be added as current study; refer to Figure-10.

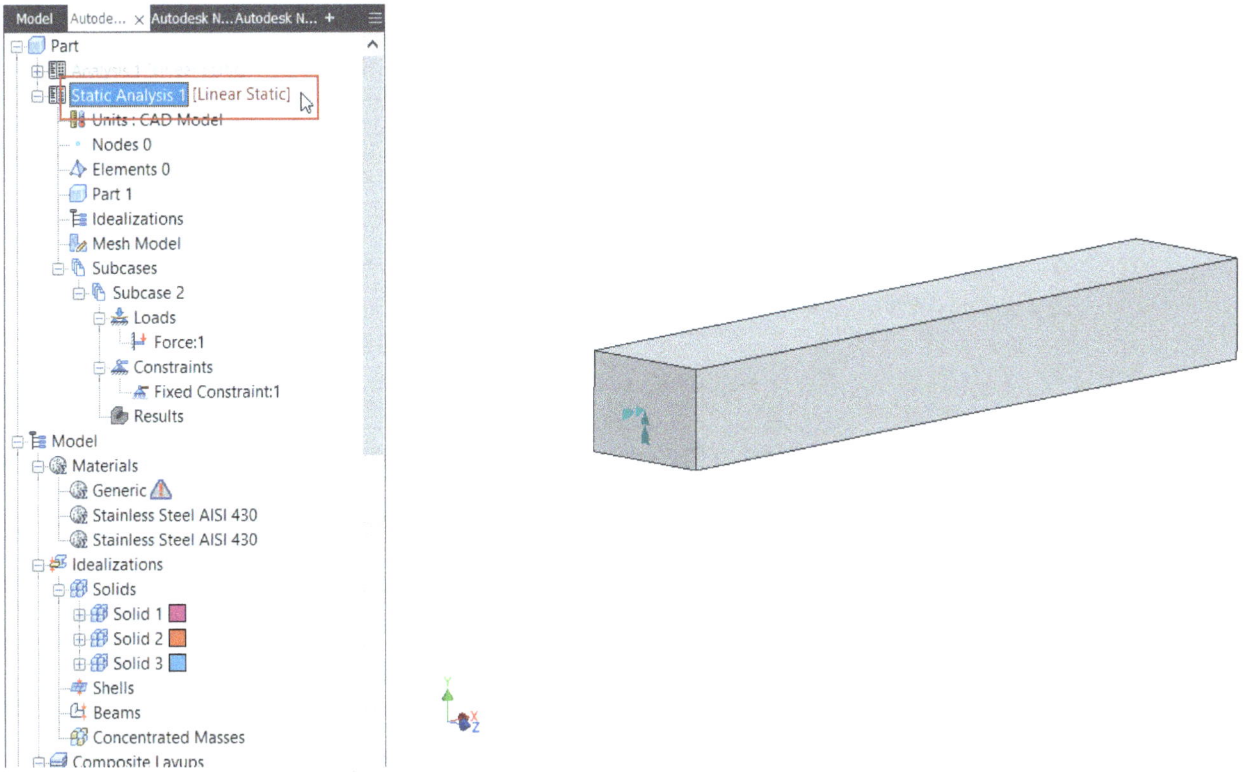

Figure-10. New analysis added by importing

APPLYING MATERIALS

The **Materials** tool is used to create, edit, and assign materials on objects under analysis. The procedure to use this tool is given next.

- Click on the **Materials** tool from the **Prepare** panel in the **Autodesk Inventor Nastran** tab. The **Material** dialog box will be displayed; refer to Figure-11.

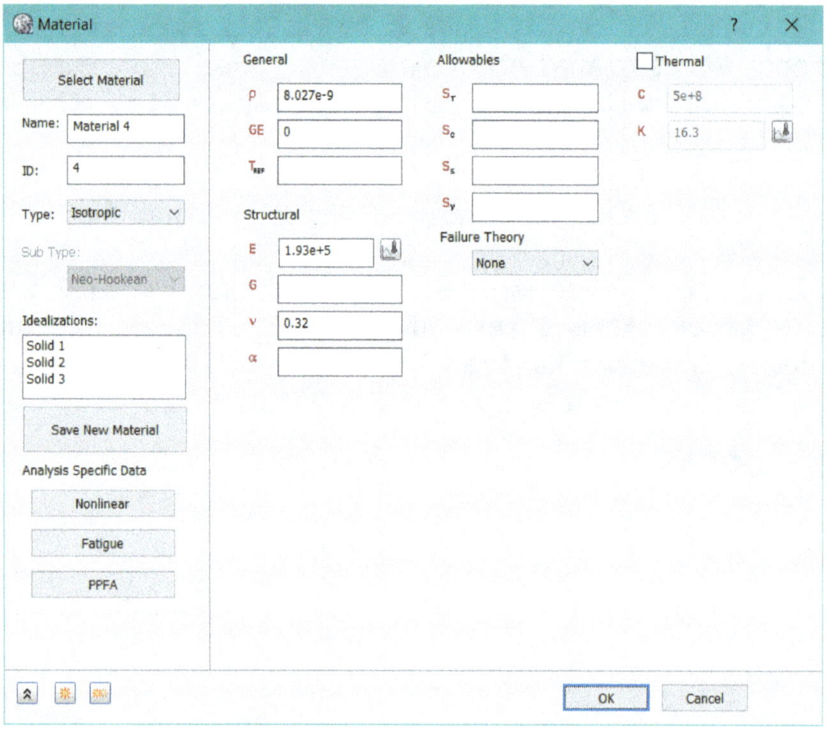

Figure-11. Material dialog box

- To select a material, click on the **Select Material** tool from the **Material** dialog box. The **Material DB** dialog box will be displayed.
- Select the desired material from the dialog box and click on the **OK** button from the **Material DB** dialog box.

Creating New Material

Although, you can use the directory of materials provided by Autodesk but there is always a requirement of new materials in engineering. If you want to add a new material which is not available by default then the procedure is given next.

- After activating the **Materials** tool, specify the desired parameters in the edit boxes of the dialog box like name of material, general parameters, structural parameters, allowables, and so on.
- Select the desired option from the **Type** drop-down.
- The options to specify analysis specific properties are available in the **Analysis Specific Data** area; like Nonlinear, Fatigue, and PPFA.
- Click on the **Nonlinear** button to specify material data for nonlinearity. The **Nonlinear Material Data** dialog box will be displayed; refer to Figure-12.

Figure-12. Nonlinear Material Data dialog box

- Select the **Nonlinear Elastic** radio button to apply nonlinear elastic property to material. Select the **Elasto-Plastic (Bi-Linear)** radio button to specify tangent modulus and initial yield stress values. Select the **Plastic** radio button to apply plastic nonlinearity to the material. After setting desired parameters, click on the **OK** button.
- Click on the **Fatigue** button from the **Analysis Specific Data** area. The **Fatigue** dialog box will be displayed; refer to Figure-13.

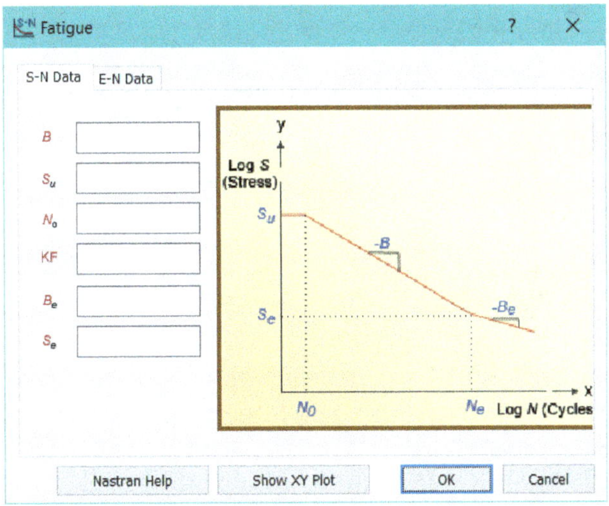

Figure-13. Fatigue dialog box

- Specify the desired parameters in the edit boxes of the dialog box and click **OK** button to define S-N Data. Click on the **E-N Data** tab from the dialog box if you want to define E-N data parameters for fatigue.
- Click on the **PPFA** button from the **Analysis Specific Data** area to create progressive ply failure analysis data. The **PPFA** dialog box will be displayed; refer to Figure-14.

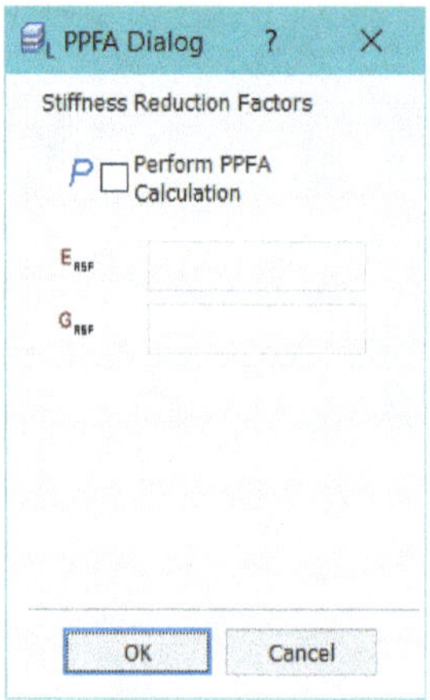

Figure-14. PPFA dialog box

- Select the **Perform PPFA Calculation** check box to specify Young's modulus and Shear modulus reduction scale factors in respective edit boxes. After setting the values, click on the **OK** button.
- Click on the **Save New Material** button to save the newly created material. The **Save As** dialog box will be displayed; refer to Figure-15.

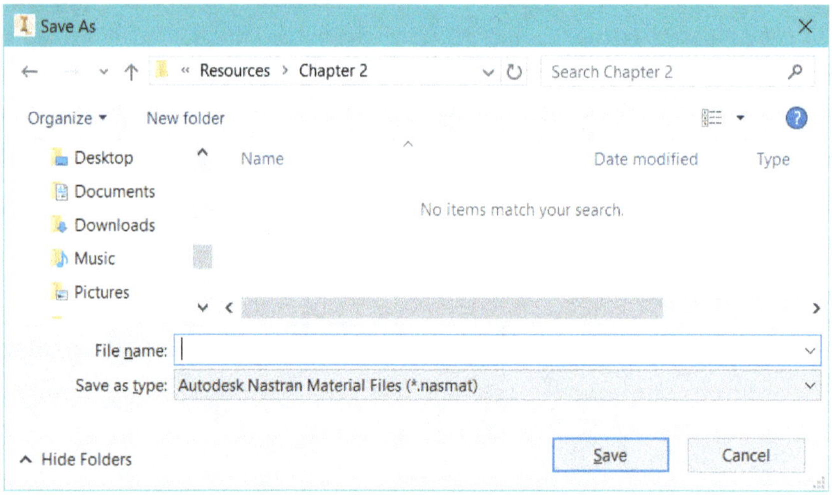

Figure-15. Save As dialog box

- Specify the desired name of material property and click on the **Save** button. The material will be created.
- After setting desired parameters, click on the **OK** button from the dialog box to exit.

IDEALIZATION

Idealization is used to idealize the model based on simple assumptions like assuming shell elements as quadrilaterals. The procedure to perform idealization is given next.

- Click on the **Idealizations** tool from the **Prepare** panel in the **Autodesk Inventor Nastran** tab of the **Ribbon**. The **Idealizations** dialog box will be displayed; refer to Figure-16.

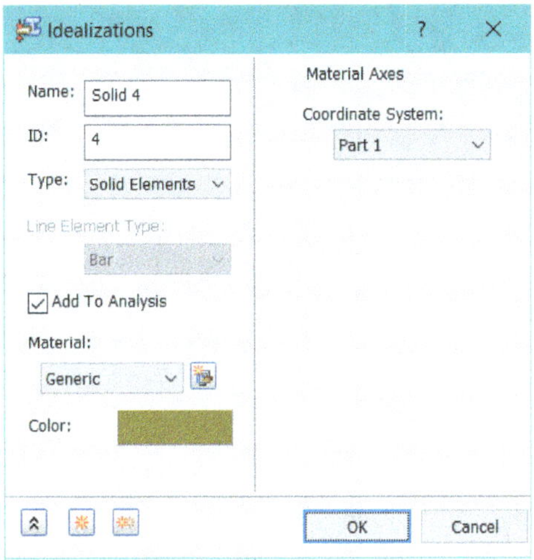

Figure-16. Idealizations dialog box

- Specify the desired name for idealization in the **Name** edit box and specify id number in the **ID** edit box to give identification code.
- From the **Type** drop-down, select the desired type of idealization to be performed on the model.
- Select the **Add To Analysis** check box to add idealization to current analysis.
- From the **Material** drop-down, select the desired material.
- Click on the colored button for **Color** field and select the desired color from the **Color** dialog box.
- To specify coordinate system for material, select the desired coordinate system from the **Coordinate System** drop-down.
- Click on the **OK** button from the dialog box to set idealization.

APPLYING CONNECTIONS BETWEEN ASSEMBLY COMPONENTS

The **Connectors** tool is used to apply connections between various components of the assembly. The procedure to use this tool is given next.

- Click on the **Connectors** tool from the **Prepare** panel of the **Autodesk Inventor Nastran** tab in the **Ribbon**. The **Connector** dialog box will be displayed; refer to Figure-17.

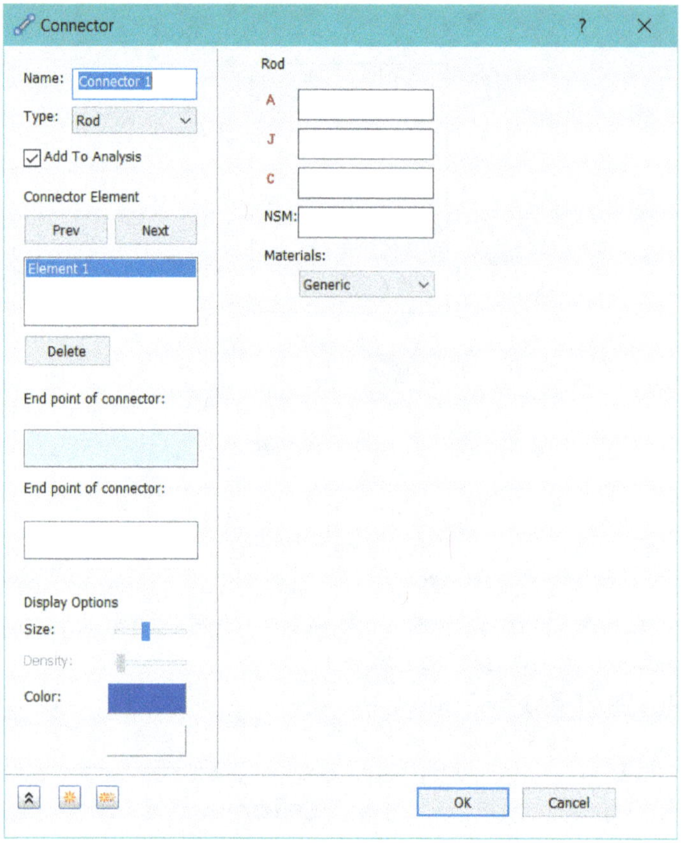

Figure-17. Connector dialog box

There are five different types of connections; Rod, Cable, Spring, Rigid Body, and Bolt. The procedures to apply these connections are given next.

Applying Rod Connection

- Select the **Rod** option from the **Type** drop-down. The **Connectors** dialog box will be displayed as shown in Figure-17.
- Click in the first **End point of connector** selection box and select the point on one end face connector.
- Click in the second **End point of connector** selection box and select the point on another end face of connector.
- Specify the desired values in different edit boxes of the **Rod** area and click on the **OK** button to apply connector.

Applying Cable Connection

- Select the **Cable** option from the **Type** drop-down to create a cable connection between selected points for analysis. The options in the dialog box will be displayed as shown in Figure-18.

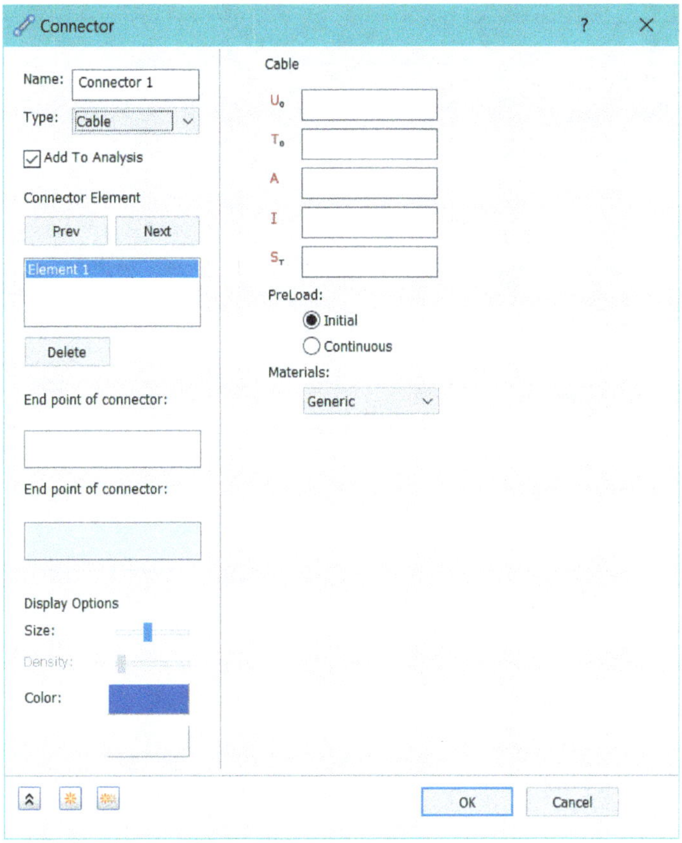

Figure-18. Connector dialog box with Cable options

- Click in the **End point of connector** selection boxes and select the desired points on the model as discussed for rod connection.
- Specify the desired values in **Cable** area of the dialog box like cable slack, cable tension, area, and so on.

Applying Spring Connection

The spring connection is used to define a connection that represents spring of specified damping and stiffness coefficient between selected points. The procedure to use this option is given next.

- Select the **Spring** option from the **Type** drop-down in **Connector** dialog box. The options in the dialog box will be displayed as shown in Figure-19.

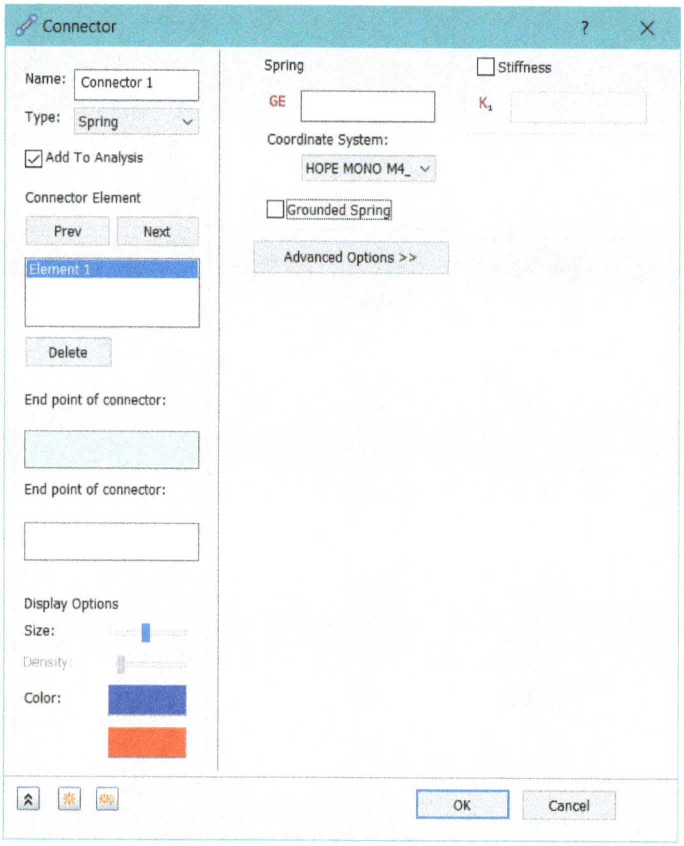

Figure-19. Connector dialog box with Spring options

- Select the connector points as discussed earlier.
- Specify the value of damping coefficient in **GE** edit box.
- Select the **Stiffness** check box and specify desired value of stiffness in the edit box below it.
- Select the **Grounded Spring** check box if one end of spring is grounded.
- Click on the **Advanced Options** button to define recovery coefficient and vector components of stiffness & damping. The options will be displayed as shown in Figure-20.

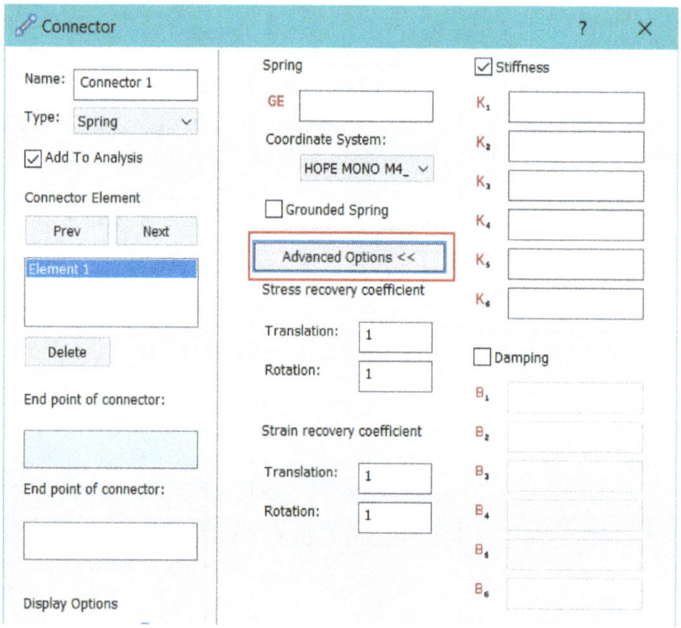

Figure-20. Advanced options for spring connection

- Specify the desired parameters for advanced options.

Applying Rigid Body Connection

Rigid body connection is used to make selected component rigid so that the object do not deform or deflect. The procedure to apply rigid connection is given next.

- Select the **Rigid Body** option from the **Type** drop-down in the **Connector** dialog box. The options in the dialog box will be displayed as shown in Figure-21.

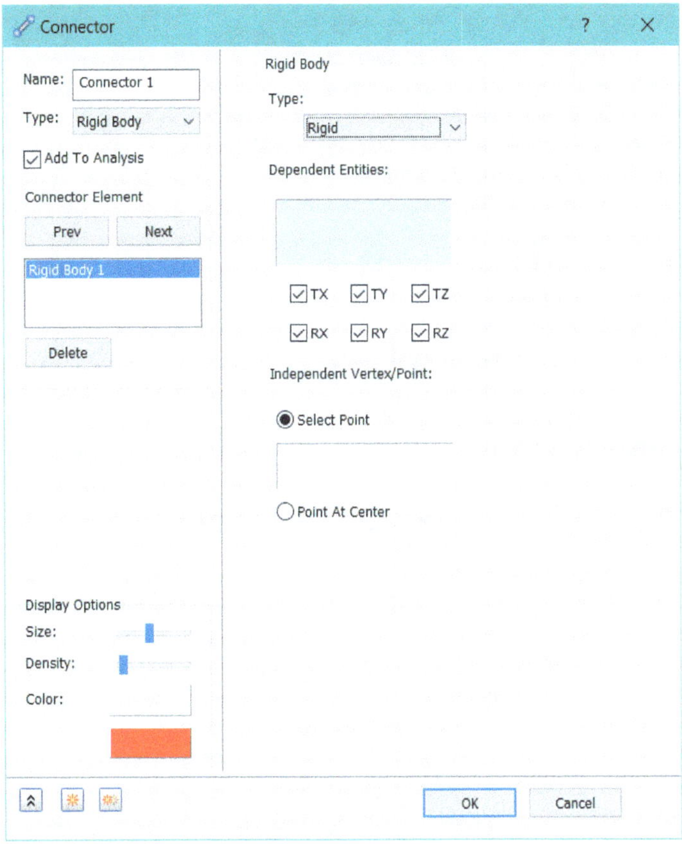

Figure-21. Connector dialog box with Rigid Body options

- Select the faces/edges/vertices that you want to be rigid.
- Click on the **Select Point** radio button to specify the center of rigid body at desired location.
- If you want the rigid body center to be at the center of selected faces/edges/vertices then select the **Point At Center** radio button.
- Set the other parameters as required and click on the **OK** button.

Applying Bolt Connection

- Select the **Bolt** option from the **Type** drop-down in **Connector** dialog box to create bolted connection. The options in the dialog box will be displayed as shown in Figure-22.
- Select the **Bolt** radio button to create a bolt with nut connection or select the **Cap Screw** radio button to create a screw connection. The options will be modified accordingly in the dialog box.

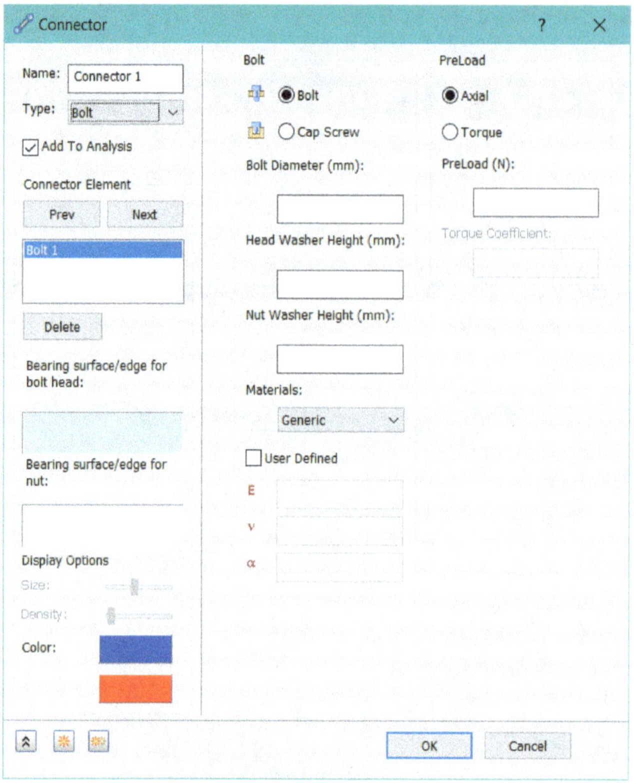

Figure-22. Connector dialog box with Bolt options

- Click in the **Bearing surface/edge for bolt head** selection box and select the edge/face on which bolt head will rest.
- Click in the **Bearing surface/edge for nut** selection box and select the edge/face on which nut will rest. Preview of the bolt will be displayed; refer to Figure-23 for bolted connection and Figure-24 for cap screw connection.

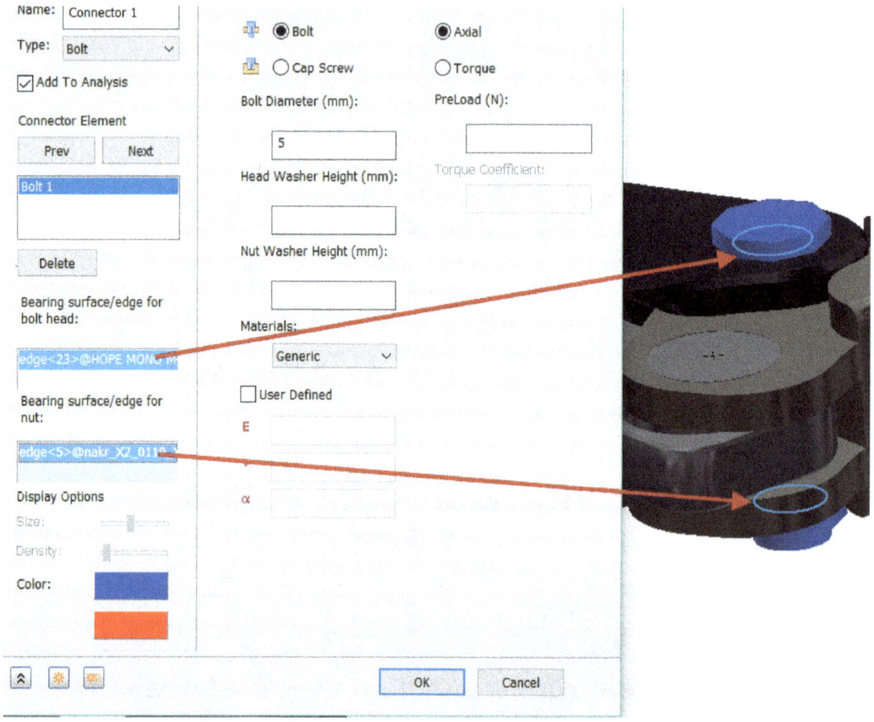

Figure-23. Edges selected for bolt

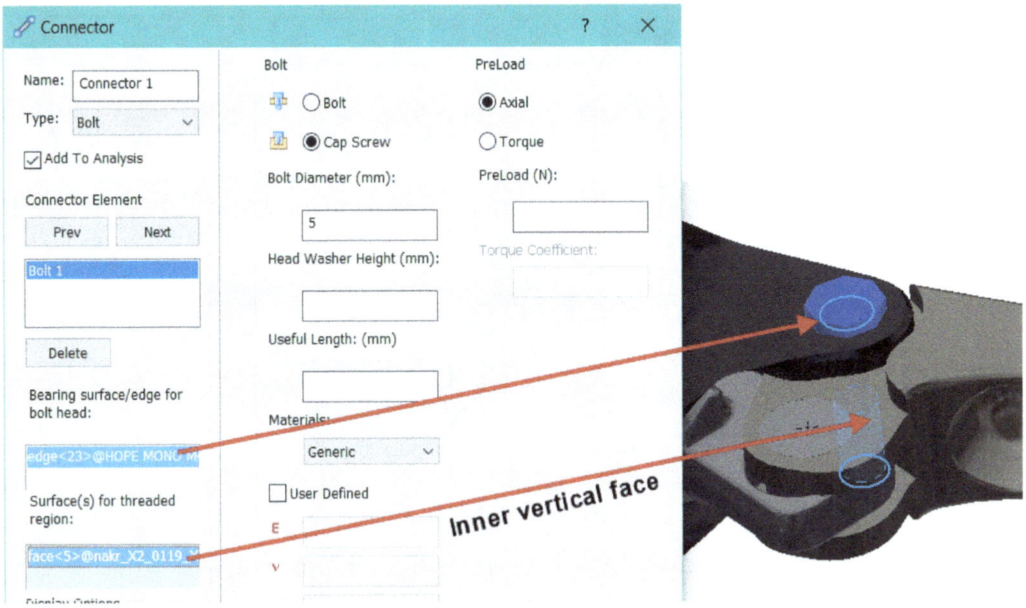

Figure-24. Edge and face selected for screw

- Specify the desired dimensions and parameters for the bolt/screw, click on the **OK** button.

OFFSETTING SURFACE FOR CREATING SHELL COMPONENTS

The **Offset Surfaces** tool in the **Prepare** panel is used to offset selected faces of the part to form shell component. The procedure to use this tool is given next.

- Click on the **Offset Surfaces** tool from the **Prepare** panel in the **Autodesk Inventor Nastran** tab of **Ribbon**. The **Offset Surface** dialog box will be displayed; refer to Figure-25.

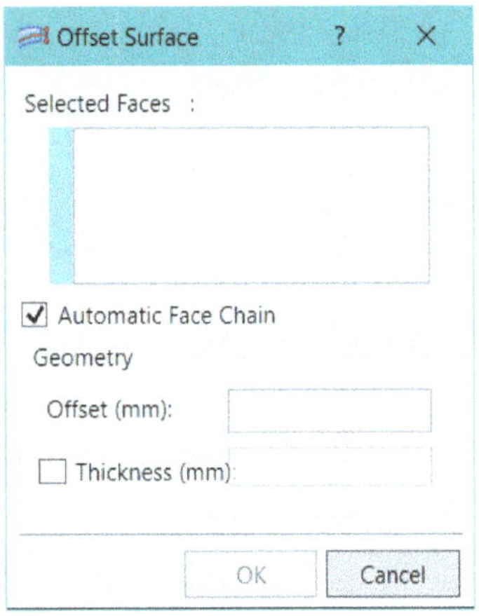

Figure-25. Offset Surface dialog box

- Select the faces that you want to offset for creating shell objects; refer to Figure-26.

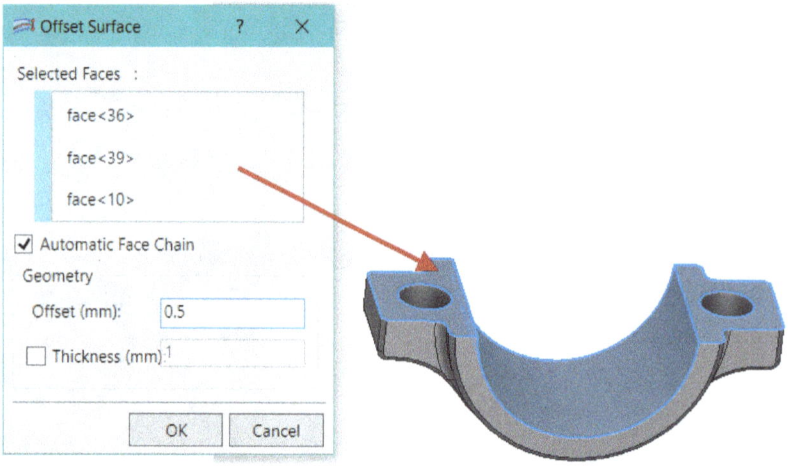

Figure-26. Faces selected

- Specify the desired value of offset in the **Offset(mm)** edit box.
- Select the **Thickness** check box to define thickness of shell component and specify the desired value in adjacent edit box.
- After setting desired parameters, click on the **OK** button to create shell component; refer to Figure-27.

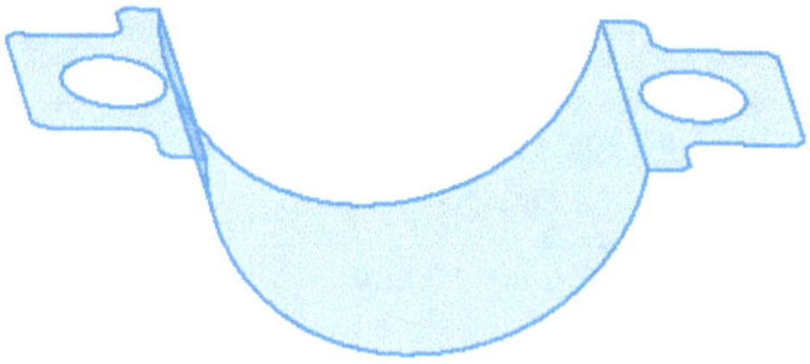

Figure-27. Shell component created

FINDING THIN BODIES

The **Find Thin Bodies** tool is used to find sections which have very low thickness as compared to rest of the model. The procedure to use this tool is given next.

- After opening the model and starting **Autodesk Inventor Nastran,** click on the **Find Thin Bodies** tool from the **Offset Surfaces** drop-down in the **Prepare** panel of the **Autodesk Inventor Nastran** tab in the **Ribbon.** If one or more thin bodies are found then **Autodesk Inventor Nastran** information box will be displayed; refer to Figure-28.

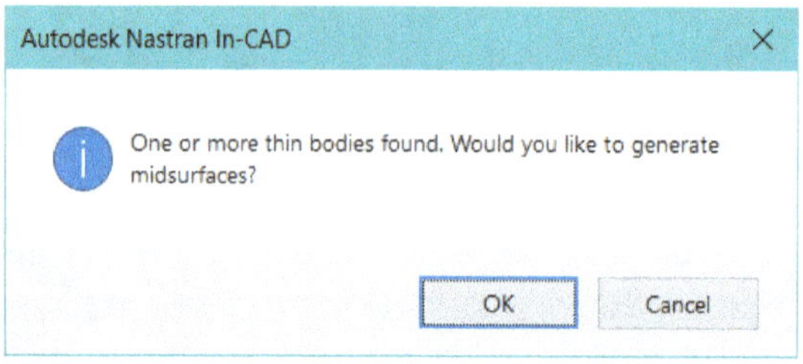

Figure-28. Autodesk Inventor Nastran information box

- If you want to generate mid surfaces then click on the **OK** button. The **Midsurface** dialog box will be displayed; refer to Figure-29.

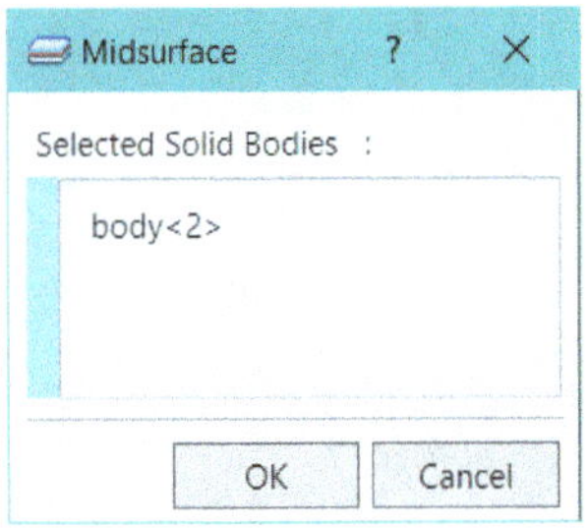

Figure-29. Midsurface dialog box

- If you want to remove any body from the list then right-click on it and select the **Delete** option from the shortcut menu.
- Click on the **OK** button from the dialog box to create the mid surfaces.

CREATING MIDSURFACE

The **Midsurfaces** tool is used to create shell objects at the mid of selected solid bodies. The procedure to use this tool is given next.

- Click on the **Midsurfaces** tool from the **Offset Surfaces** drop-down in the **Prepare** panel of the **Autodesk Inventor Nastran** tab in the **Ribbon**. The **Midsurface** dialog box will be displayed; refer to Figure-29.
- Select the bodies for which you want to create midsurface; refer to Figure-30.

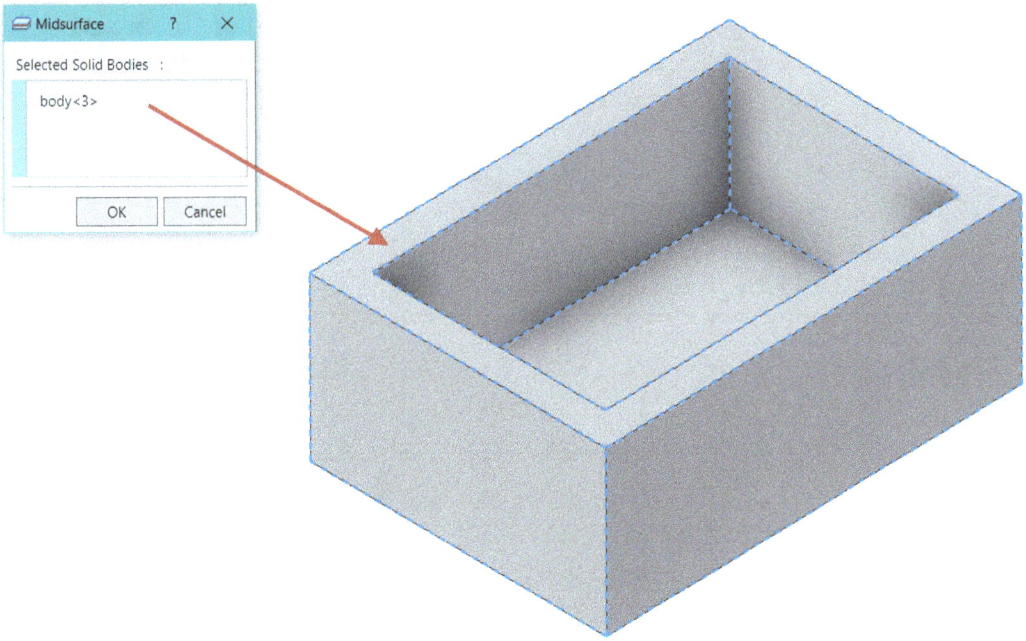

Figure-30. Body selected for midsurface

- Click on the **OK** button from the dialog box. The midsurface will be created; refer to Figure-31.

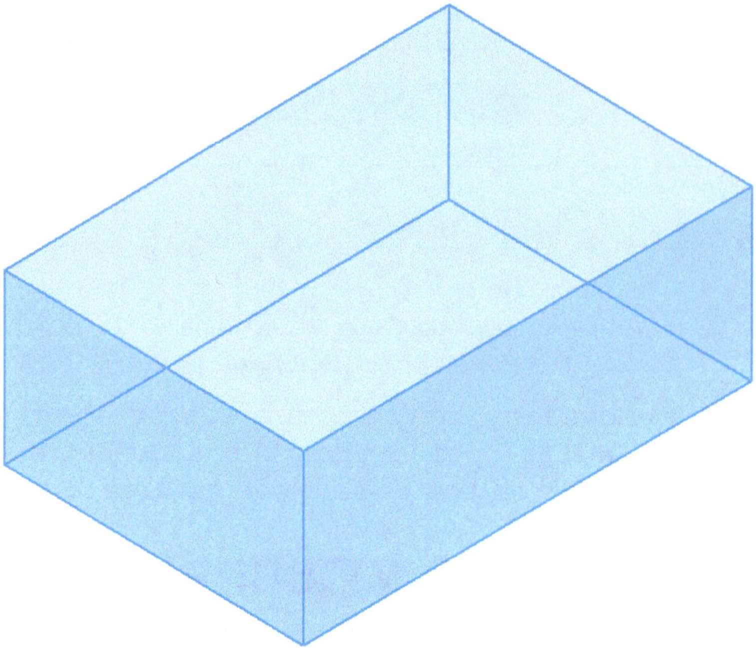

Figure-31. Midsurface generated

MANAGING STRUCTURAL MEMBERS

The **Structural Members** tool is used to manage whether the structural members in model are to be used as solids or beams in analysis. Note that structural members are generated by Frame Generator environment of Autodesk Inventor. The procedure to use this tool is given next.

- Click on the **Structural Members** tool from the **Prepare** panel of **Autodesk Inventor Nastran** tab in the **Ribbon**. The **Structural Members** dialog box will be displayed; refer to Figure-32.

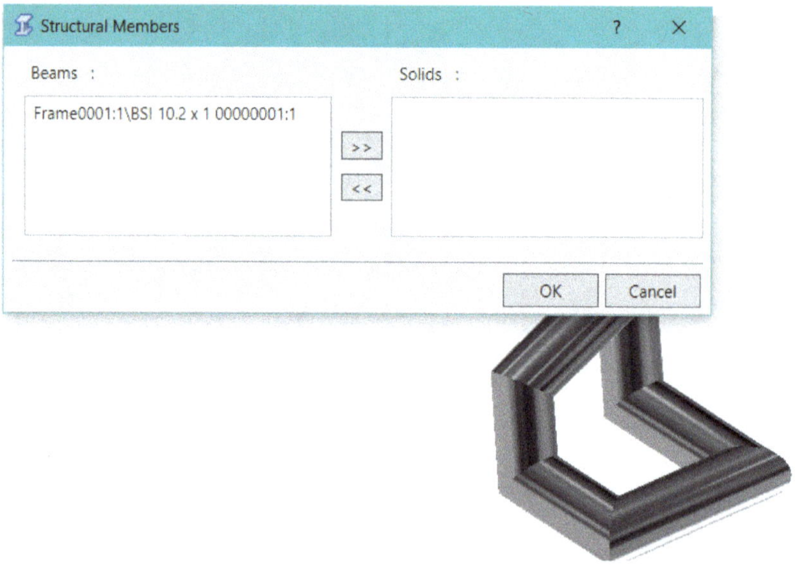

Figure-32. Structural Members dialog box

- Select the desired frame members from the **Beams** list box and click on the Forward(>>) button to use them as solid members in analysis. Similarly, select the desired frame members from the Solids list box and click on the Backward (<<) button to use them as beam members in analysis.

- After setting the desired parameter, click on the **OK** button from the dialog box.

APPLYING CONSTRAINTS

The **Constraints** tool in **Setup** panel is used to apply restrictions to translation and rotation of selected objects. The procedure to apply constraint is given next.

- Click on the **Constraints** tool from the **Setup** panel in the **Autodesk Inventor Nastran** tab of the **Ribbon**. The **Constraint** dialog box will be displayed; refer to Figure-33.

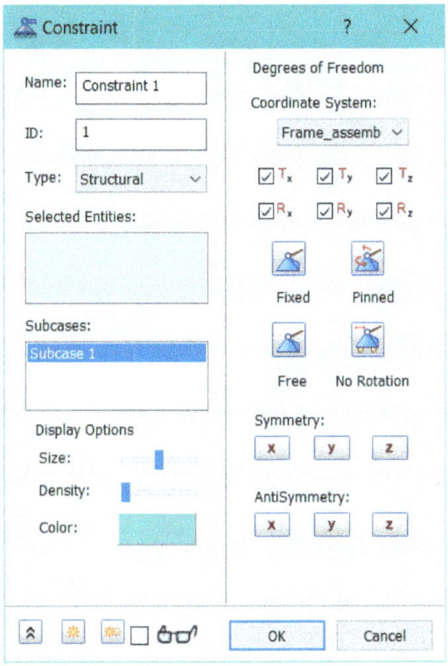

Figure-33. Constraint dialog box

- Select the desired type of constraint from the **Type** drop-down. The options will be modified accordingly in the dialog box. If you have selected **Thermal** option from the **Type** drop-down then you will be asked to specify temperature on selected entities.
- Select the faces/edges/vertices to which you want to apply constraint.
- Select the desired check boxes from T_x, T_y, T_z, R_x, R_y, and R_z from the **Degrees of Freedom** area. Here T designates translation along subscripted direction and R designates rotation along subscripted direction. You can also use the **Fixed** button to restrict translation and rotation of selected objects, **Pinned** button to restrict only translation of selected objects, **Free** button to allow free movement of selected objects, and **No Rotation** button to restrict only rotation of selected objects.
- Select the desired button from the **Symmetry** area to make the object symmetric along axis of selected button. Similarly, you can use the buttons in **AntiSymmetry** area.
- Set the desired option from the **Display Options** area like display size, display density and color of constraint.
- Click on the **New** button at the bottom of the dialog box to create another constraint.

- After setting desired parameters, click on the **OK** button.

APPLYING LOADS

The **Loads** tool is used to apply different types of loads on selected faces/edges/vertices. The procedure to apply different types of loads is discussed next.

- Click on the **Loads** tool from the **Setup** panel in the **Autodesk Inventor Nastran** tab of **Ribbon**. The **Load** dialog box will be displayed; refer to Figure-34.

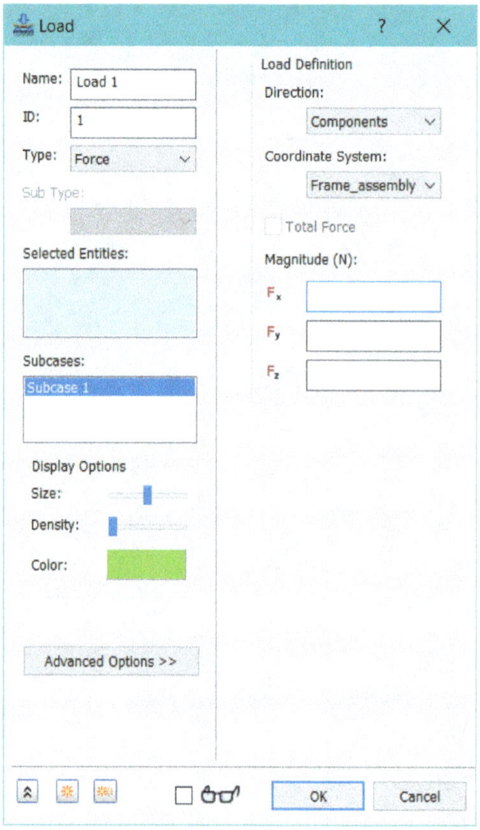

Figure-34. Load dialog box

- Specify the desired load name and id in respective edit boxes.
- Set the desired display options in **Display Options** area.

Applying Force

- Select the **Force** option from the **Type** drop-down.
- Select the faces on which you want to apply force.
- Specify different components of force in Fx, Fy, and Fz edit boxes of **Magnitude** area if **Components** option is selected in **Direction** drop-down.
- Select the **Normal to Surface** option from the **Direction** drop-down to apply force perpendicular to selected face and specify the magnitude in **Magnitude** edit box.
- Select the **Geometric Entity** option from the **Direction** drop-down to apply force along selected direction reference; refer to Figure-35.

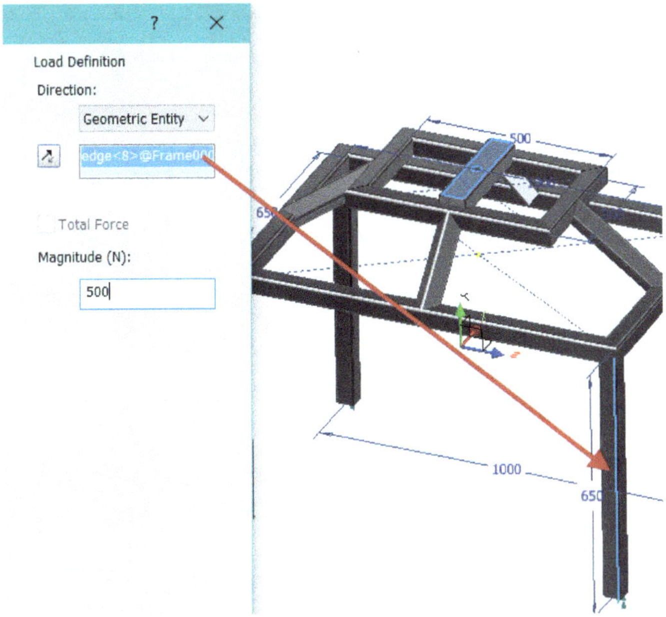

Figure-35. Edge selected for direction

- Click on the **Advanced Options** button from the dialog box to set variable load. The options of the dialog box will be displayed as shown in Figure-36.

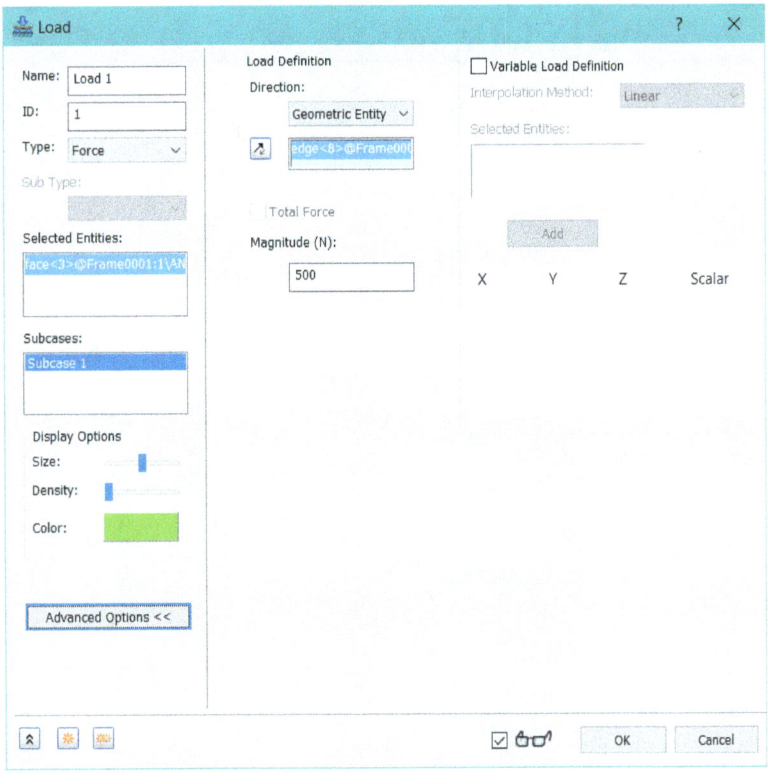

Figure-36. Advanced options for force load

- Select the **Variable Load Definition** check box to create variable loading. The options below the check box will become active.
- Select the **Linear** option if you want the load to vary linearly along selected point. Minimum four points are required for a face. Select the **Linear Equation Based** option to vary load linearly based on standard plane equation. Minimum 3 points are required for a face

load. Select the **Quadratic** option to vary load based on parabolic equation. Figure-37 shows the difference between load distributions of these options.

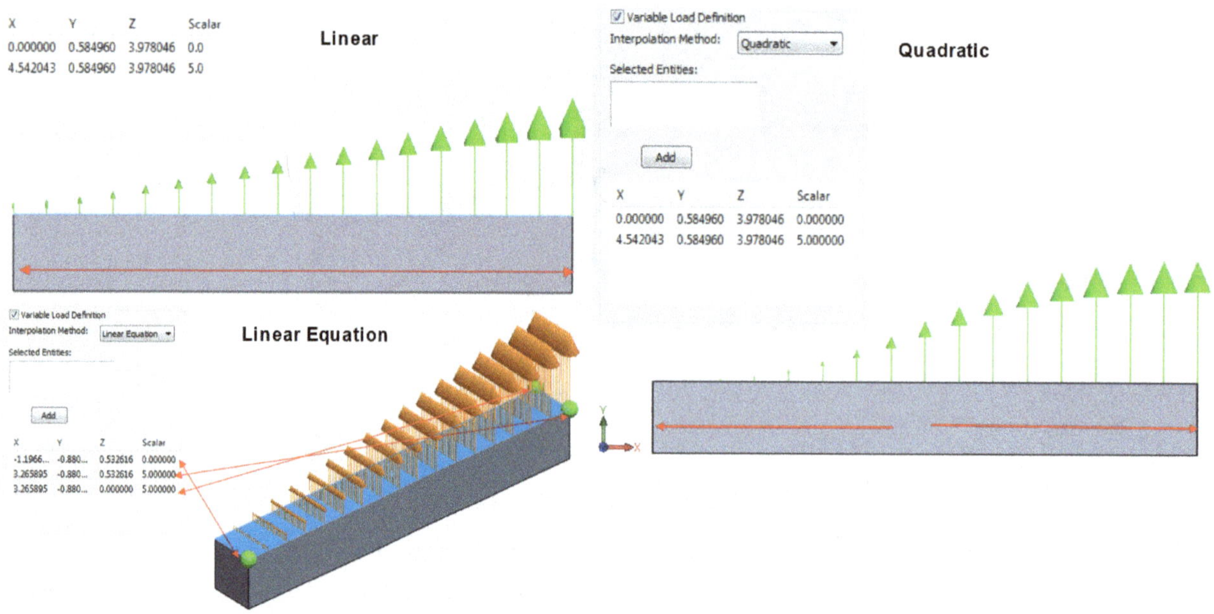

Figure-37. Variable load distribution

Applying Moment Load

- Moment is load applied on part which rotates or tends to rotate the part. After activating **Load** dialog box, select the **Moment** option from the **Type** drop-down. The options in the dialog box will be displayed as shown in Figure-38.

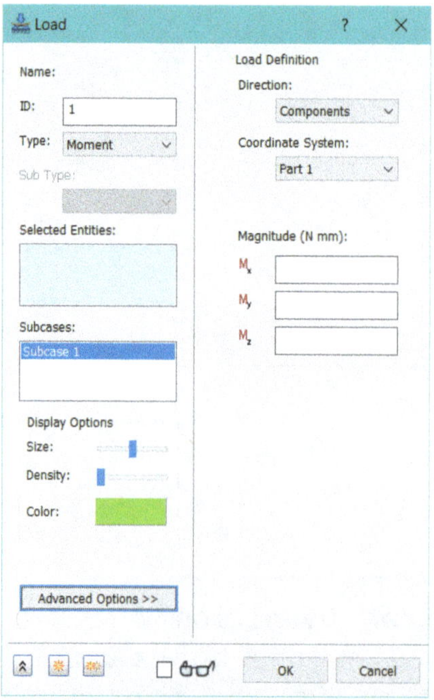

Figure-38. Load dialog box with Moment options

- Most of the options in this dialog box are same as discussed for Force option. Specify the value of moment in place of force and click on the **OK** button to apply changes.

Applying Distributed Load

Distributed load is applied when you need to specify the force per unit length of the object; refer to Figure-39. To apply distributed load, select the **Distributed Load** option from the **Type** drop-down. The options in this dialog box are same as discussed earlier. Select the desired option from the **Sub Type** drop-down. The other options are same as discussed earlier.

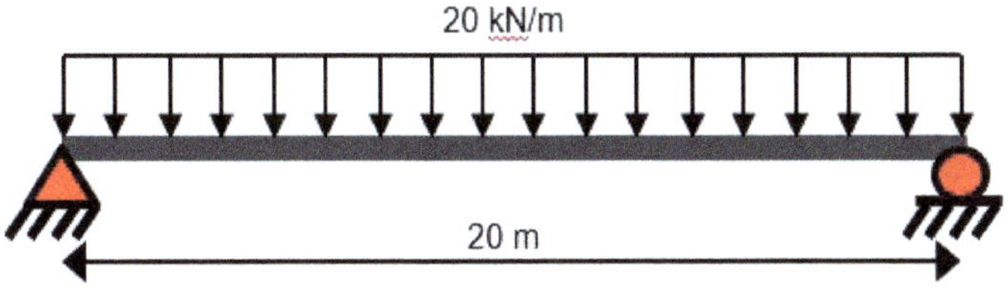

Figure-39. Distributed load

Hydrostatic Load

The hydrostatic load is applied when you need to define force exerted by fluid (at rest) on a surface. In simple words, hydrostatic force is equal to force exerted due to pressure of fluid at rest. Hydrostatic pressure is calculated as fluid density x depth of fluid. Hydrostatic force is maximum at bottom and minimum on top of wall. The procedure to apply this load is given next.

- Select the **Hydrostatic Load** option from the **Type** drop-down in the dialog box. The options will be displayed as shown in Figure-40.

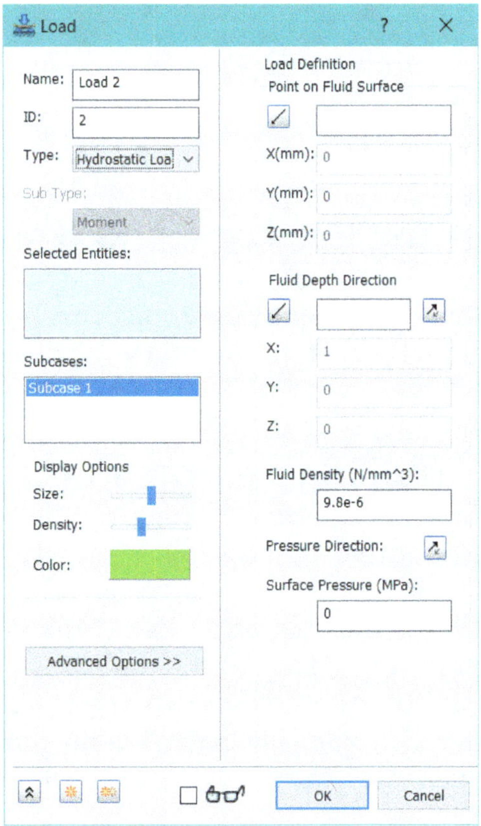

Figure-40. Options for hydrostatic load

- Select the face/surface on which you want to apply the load.
- Click on the button in **Point on Fluid Surface** area of the dialog box and select the top point upto which fluid is exerting pressure.
- Click on the button in **Fluid Depth Direction** area and select the direction reference (edge or axis).
- Specify the desired values for fluid density and surface pressure in respective fields. Preview of hydrostatic load will be displayed; refer to Figure-41.

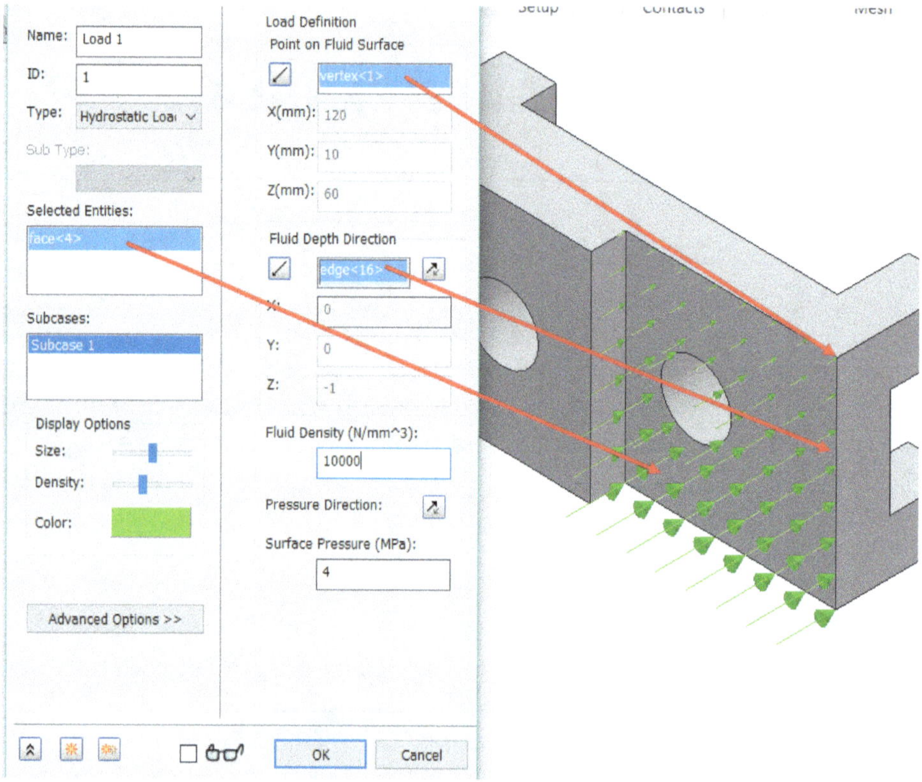

Figure-41. Preview of hydrostatic load

- Click on the **OK** button from the dialog box to create the load.

Applying Pressure and Gravity

The procedures to apply pressure and gravity are same as discussed earlier for force. Select the **Pressure** option from the **Type** drop-down to apply pressure and select the **Gravity** option from the **Type** drop-down to apply gravity load.

Applying Remote Force

The **Remote Force** option is used to apply effect of distant load at specified point. The procedure to apply remote force is given next.

- Select the **Remote Force** option from the **Type** drop-down. The options in the dialog box will be displayed as shown in Figure-42.
- Select the face on which you want to apply load.
- Click in the selection box of **Remote Force Location** area and select the point at which remote force is located; refer to Figure-43.

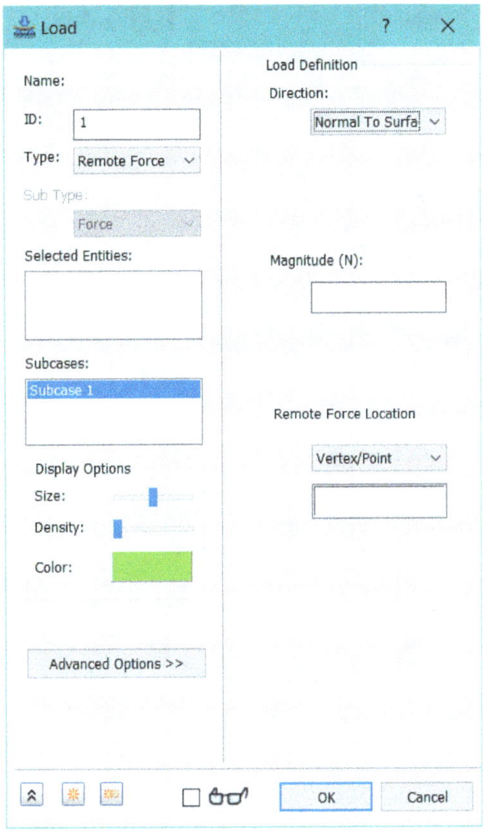

Figure-42. Load dialog box with Remote Force options

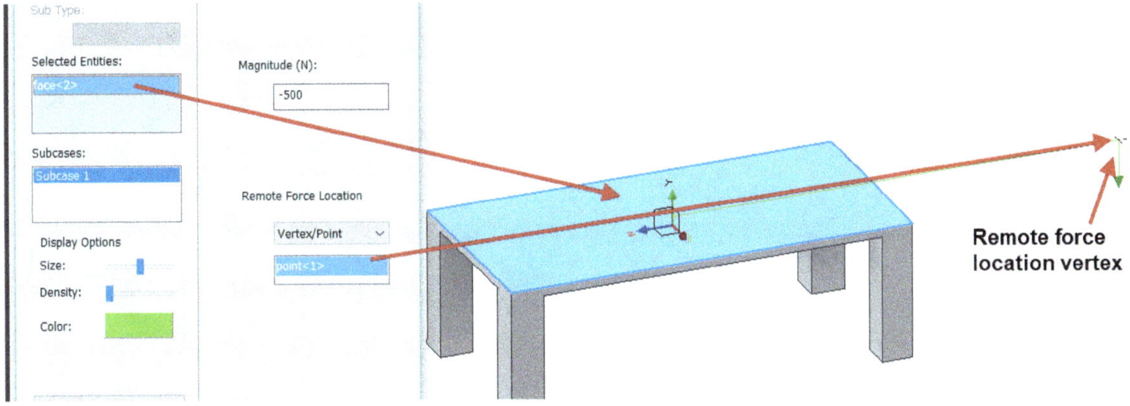

Figure-43. Applying remote force

- Specify the desired value of remote force in **Magnitude** edit box.
- The advanced options are same as discussed earlier. Click on the **OK** button from the dialog box to apply load.

Applying Bearing Load

Bearing load is used to apply radial and axial forces on selected round faces to represent bearing load. The procedure to apply bearing load is given next.

- Select the **Bearing Load** option from the **Type** drop-down after activating the **Load** dialog box. The options in this dialog box will be displayed as shown in Figure-44.

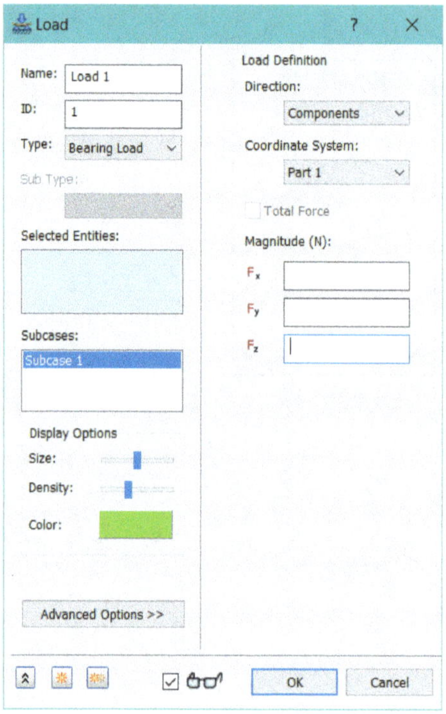

Figure-44. Load dialog box with Bearing Load options

- Select the round face on which bearing load is to be applied.
- Specify various components of bearing load in edit boxes of **Magnitude** area; refer to Figure-45.

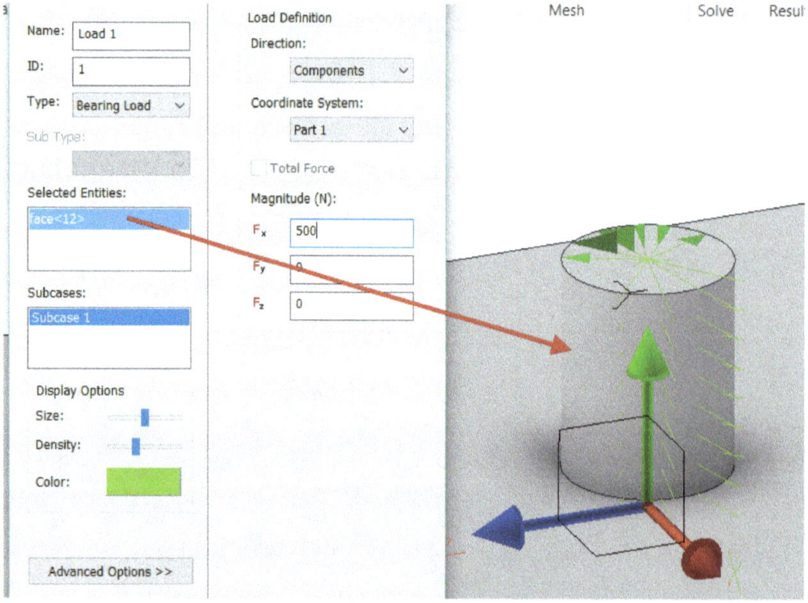

Figure-45. Applying bearing load

- After setting desired parameters, click on the **OK** button to apply bearing load.

Applying Rotational Force

Rotational force is applied to rotate the model at defined velocity and acceleration. The procedure is given next.

- Select the **Rotational Force** option from the **Type** drop-down after activating **Load** dialog box. The options in this dialog box will be displayed as shown in Figure-46.

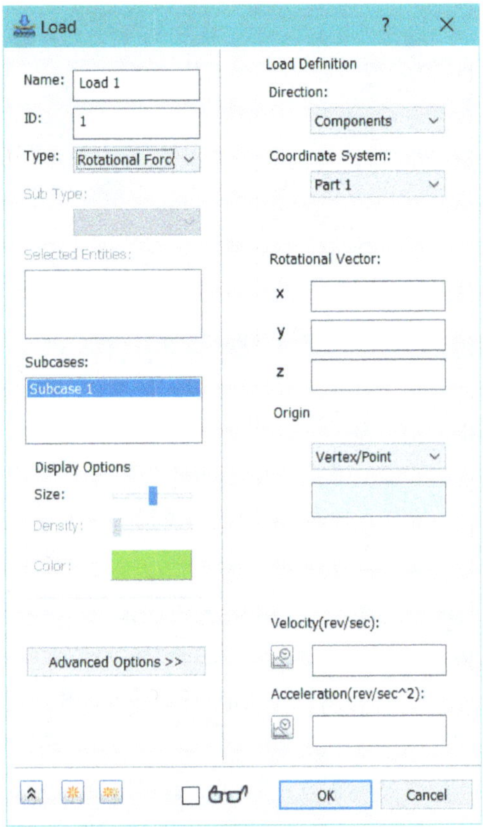

Figure-46. Load dialog box with Rotational Force options

- Click in the **Selected Entities** selection box and select the face of model on which you want to apply rotational force.
- Click in the selection box of **Origin** area and select the point on rotational axis to define origin for rotation; refer to Figure-47.
- Specify the desired value of velocity and acceleration in respective edit boxes.

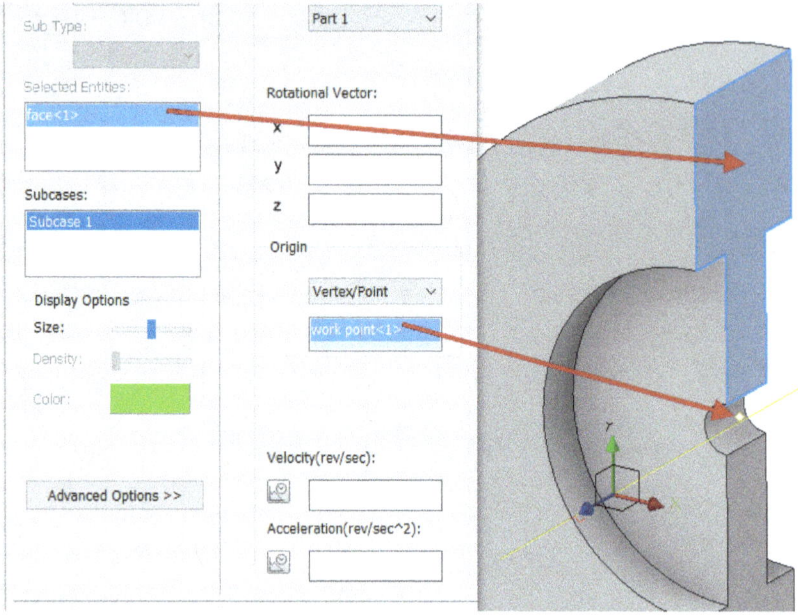

Figure-47. Applying rotational force

- Set desired parameters and click on the **OK** button.

Applying Enforced Motion Load

The **Enforced Motion** option is used to apply specified translation or rotation. The procedure to do so is given next.

- Select the **Enforced Motion** option from the **Type** drop-down after activating the **Load** dialog box. The options related to enforced motion will be displayed.
- Select the desired option from **Sub Type** drop-down.
- Select the face on which you want to apply translation or rotation and specify the related value in **Magnitude** edit box.
- After setting desired values, click on the **OK** button from the dialog box.

Applying Initial Condition

The **Initial Condition** option is used to specify initial temperature for the study. The initial condition is applicable for nonlinear and dynamic analyses. The procedure to use this option is given next.

- Select the **Initial Condition** option from the **Type** drop-down after activating the **Load** dialog box. The options will be displayed as shown in Figure-48.
- Specify the desired temperature in the **Temperature** edit box.
- After setting desired values, click on the **OK** button.

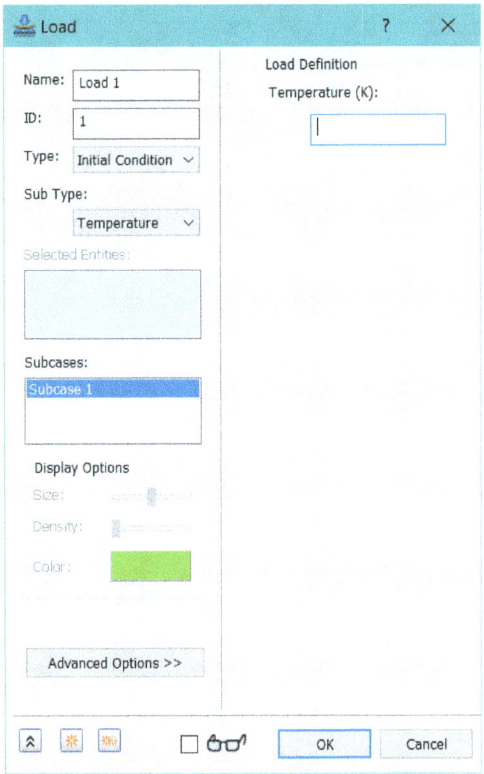

Figure-48. Load dialog box with Initial Condition options

Applying Body Temperature

The **Body Temperature** option is used to set desired temperature for the model. To set body temperature, select the **Body Temperature** option from the **Type** drop-down of **Load** dialog box. The options for body temperature are same as for initial condition of temperature. After setting desired parameters, click on the **OK** button.

Applying Temperature

The **Temperature** option is used to specify desired temperature to selected faces/edges/vertices. The procedure to use this option is given next.

- After activating the **Load** dialog box, select the **Temperature** option from the **Type** drop-down. The options in the dialog box will be displayed as shown in Figure-49.
- Select the desired faces/edges/vertices to which you want to apply temperature.
- Specify the desired temperature value in the **Temperature** edit box.
- Set the other parameters as required and click on the **OK** button.

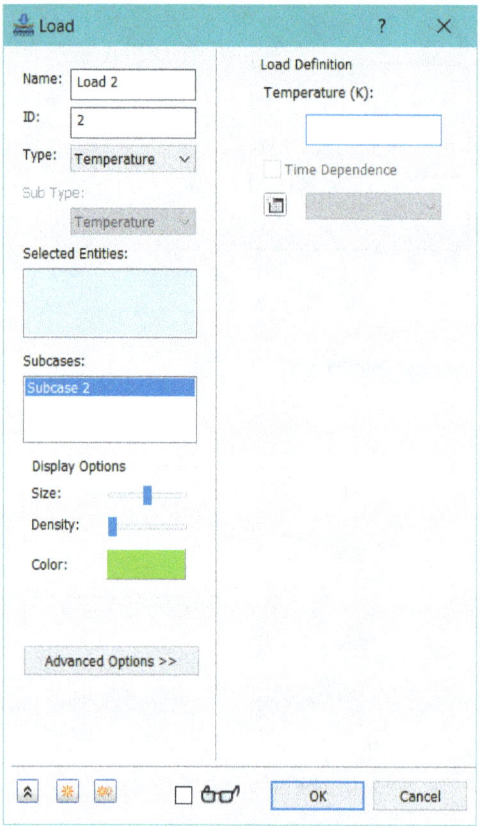

Figure-49. Load dialog box with Temperature options

Applying Thermal Convection

The **Thermal Convection** option is used to apply thermal convection coefficient to selected faces/edges/vertices. The procedure to use this tool is given next.

- Select the **Thermal Convection** option from the **Type** drop-down in the **Load** dialog box. The options in the dialog box will be displayed as shown in Figure-50.
- Select the faces/edges/vertices on which you want to specify heat convection source.
- Specify the surrounding environment temperature in **Ambient Temperature** edit box.
- Specify the convection coefficient value in **Convection Coefficient** edit box.
- If you want to make convection rate temperature dependent then select the **Temperature Dependence** check box. The table options below it will become active. Click on the **Define New Table** button to define convection coefficient at different temperatures. The **Table Data** dialog box will be displayed; refer to Figure-51.

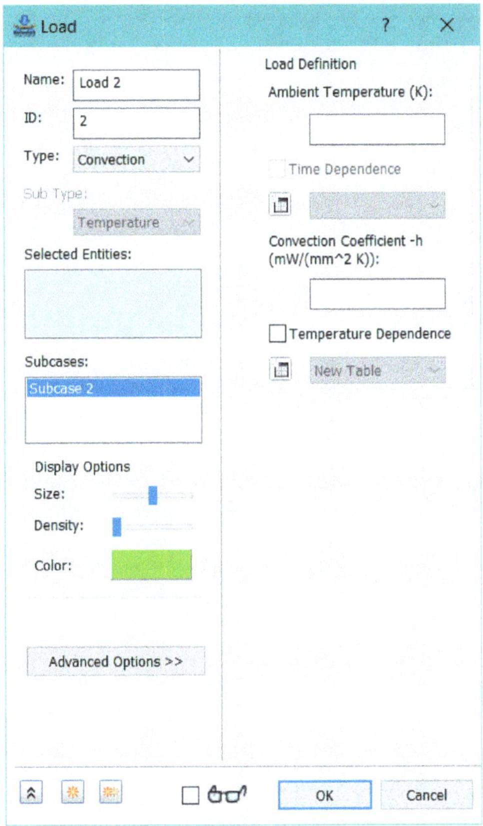

Figure-50. Load dialog box with Convection options

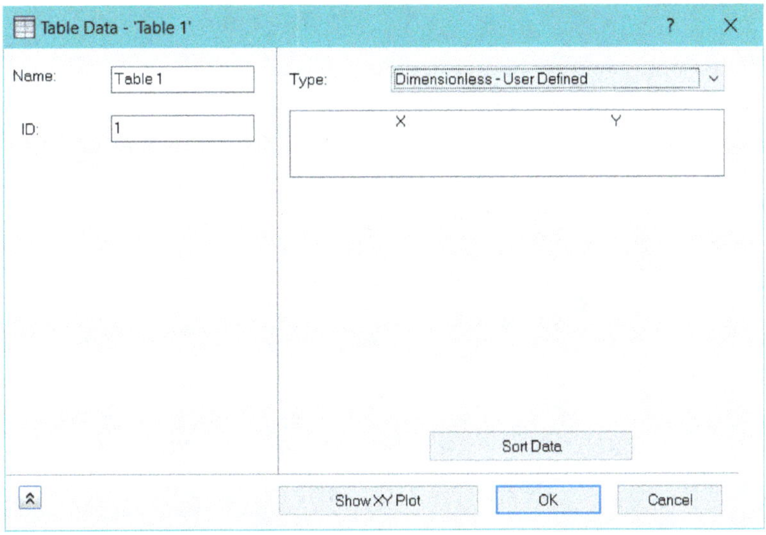

Figure-51. Table Data dialog box

- Select the **Convection Coefficient vs. Temperature** option from the **Type** drop-down to specify convection coefficients at different temperatures.
- Double-click in the fields and specify the desired values; refer to Figure-52.

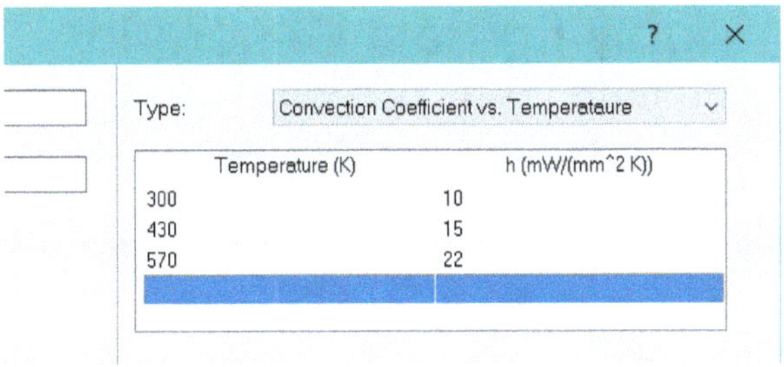

Figure-52. Convection rates specified for temperatures

- If you want to check the plot of convection rate v/s temperature then click on the **Show XY Plot** button. After setting the desired parameters, click on the **OK** button from the **Table Data** dialog box to apply convection coefficient. The Load dialog box will be displayed again.
- After setting desired parameters, click on the **OK** button from the **Load** dialog box.

Applying Thermal Radiation

- After activating the **Load** dialog box, select the **Radiation** option from the **Type** drop-down. The options in the dialog box will be displayed as shown in Figure-53.

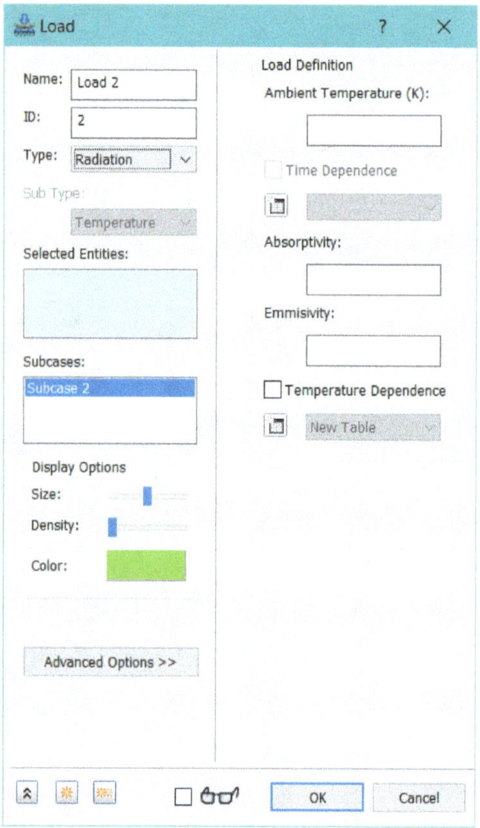

Figure-53. Load dialog box with Radiation options

- Select the faces/edges/vertices on which you want to apply radiation heat.
- Specify the desired values for temperature, absorptivity, and emmisivity in respective edit boxes of the dialog box.
- After setting desired parameters, click on the **OK** button.

Applying Heat Generation

The **Heat Generation** option is used to apply heat generation at selected faces/edges/vertices. The procedure to use this option is given next.

- Select the **Heat Generation** option from the **Type** drop-down. The options in the dialog box will be displayed as shown in Figure-54.

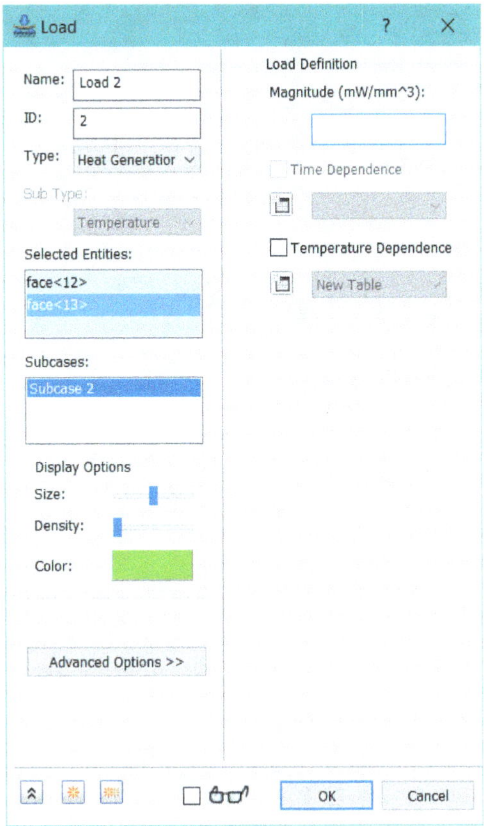

Figure-54. Load dialog box with Heat Generation options

- Select the faces/edges/vertices to which you want to specify heat generation rate.
- Specify the desired value of heat generation in the **Magnitude** edit box.
- Set the other options as discussed earlier and click on the **OK** button to apply load.

Applying Heat Flux

Heat Flux is the heat flow rate intensity per unit of area per unit time. The procedure to apply heat flux is given next.

- After activating **Load** dialog box, select the **Heat Flux** option from the **Type** drop-down. The options in the **Load** dialog box will be displayed as shown in Figure-55.

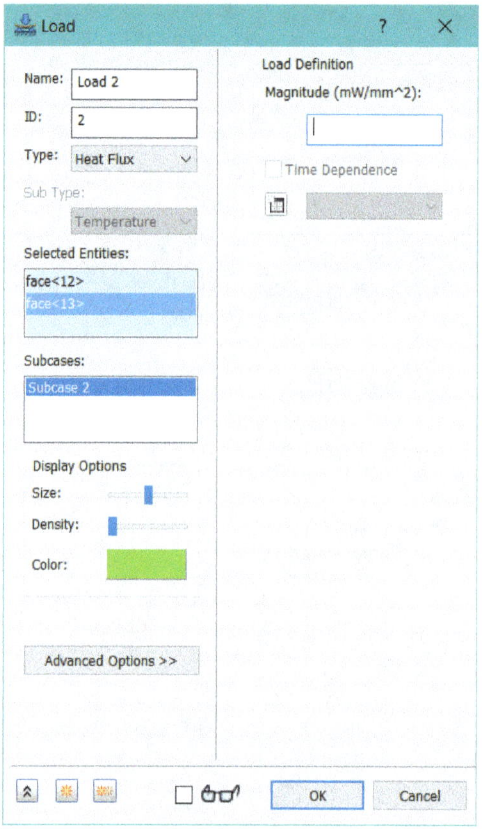

Figure-55. Load dialog box with Heat Flux options

- Select the faces/edges/vertices to which you want to apply heat flux and specify the desired heat flux value in **Magnitude** edit box.
- After setting desired values, click on the **OK** button from the dialog box.

Applying Loads based on Analysis Results

- Select the **From Output** option from the **Type** drop-down. The options of **Loads** dialog box will be displayed as shown in Figure-56.
- Click on the **Browse** button next to **Results File** edit box. The **Open** dialog box will be displayed; refer to Figure-57.
- Select the desired result file (.fno or .cfdst) and click on the **Open** button. The nodal load and output set parameters will be displayed in respective drop-downs.
- Set the desired parameters and click on the **OK** button to apply loads.

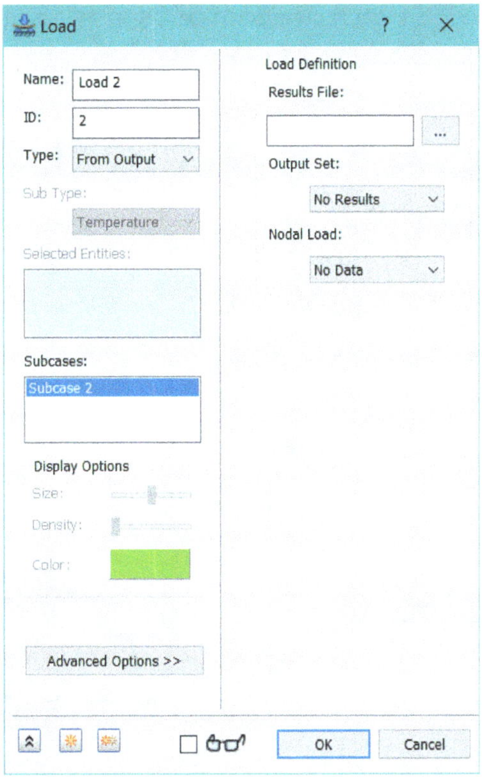

Figure-56. Load dialog box with From Output options

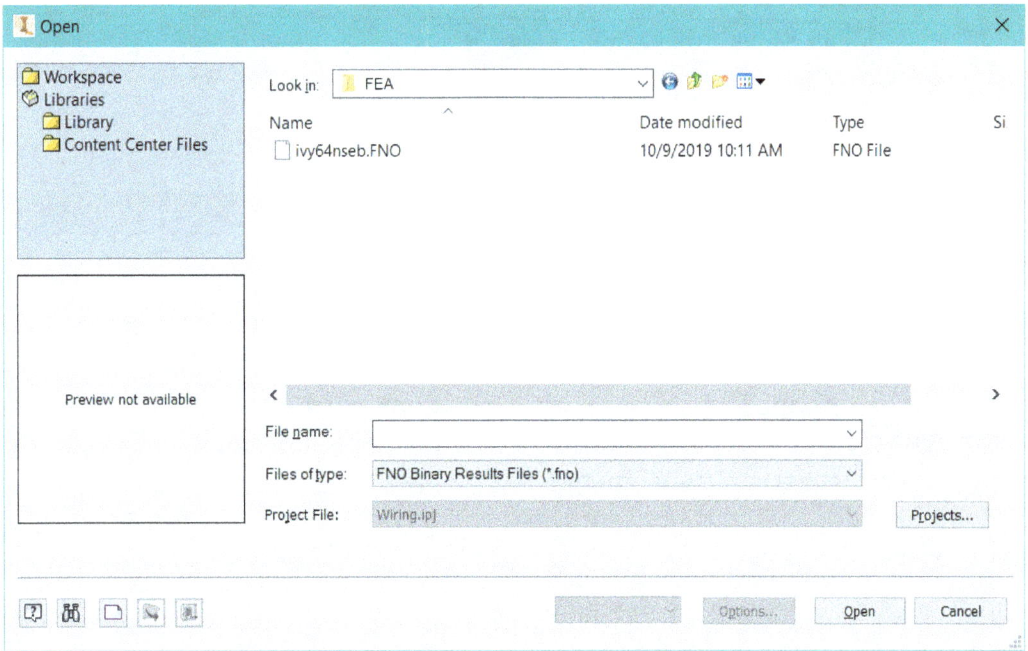

Figure-57. Open dialog box

DEFINING CONTACT SETS

There are three tools to define contact sets viz. **Auto**, **Manual**, and **Solver**. The procedures to use these tools are discussed next.

Applying Automatic Contact Sets

The **Auto** tool is used to automatically set contacts in the assembly. The procedure to set contacts automatically is given next.

- Click on the **Auto** tool from the **Contacts** panel in the **Autodesk Inventor Nastran** tab of **Ribbon**. The information box will be displayed; refer to Figure-58.

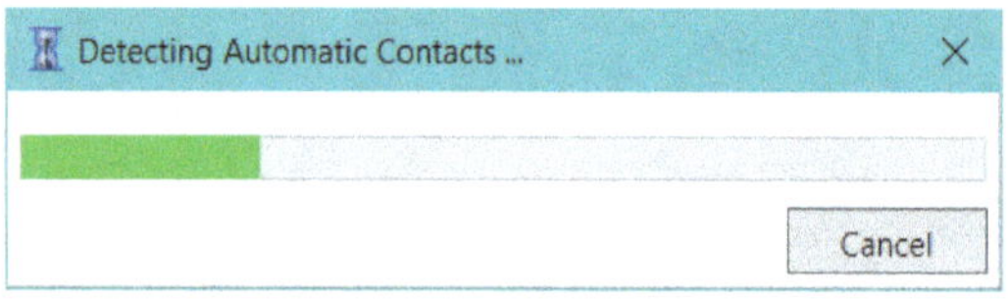

Figure-58. Detecting Automatic Contacts information box

- Once the detection is complete, the contacts will be applied automatically; refer to Figure-59.

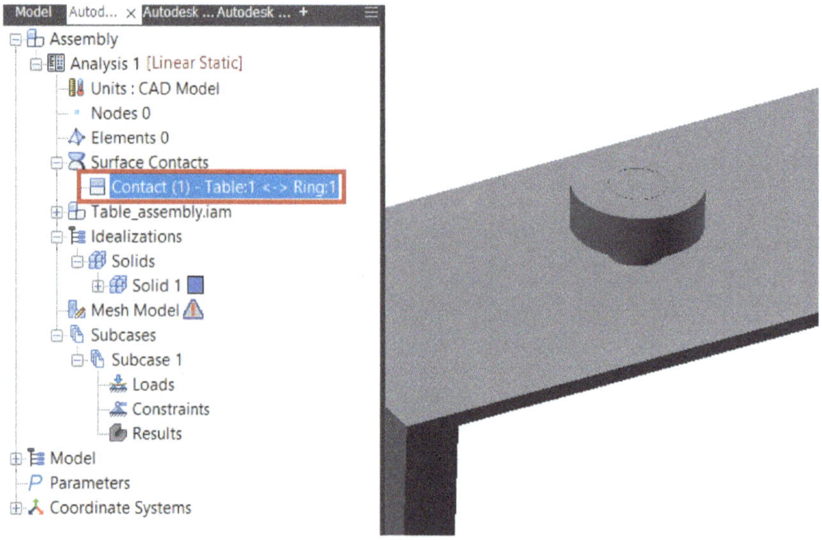

Figure-59. Contact applied automatically

Applying Manual Contacts

The **Manual** tool in **Contacts** panel is used to apply different types of contacts manually to assembly components. The procedure to use this tool is given next.

- Click on the **Manual** tool from the **Contacts** panel in the **Autodesk Inventor Nastran** tab of **Ribbon**. The **Surface Contact** dialog box will be displayed; refer to Figure-60.
- Specify the desired name and id for contact in **Name** and **ID** edit boxes respectively.

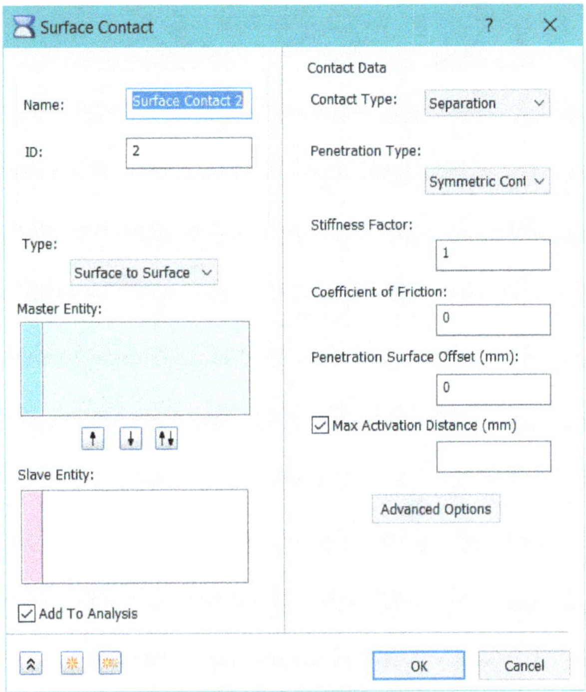

Figure-60. Surface Contact dialog box

- Select the desired contact type from the **Contact Type** drop-down. The options in this drop-down are **Separation**, **Bonded**, **Sliding/No Separation**, **Separation/No Sliding**, **Offset Bonded**, **Shrink Fit/Sliding**, **Shrink Fit/No Sliding**, and **Disable**. Select the **Separation** option if there is one object placed over another without any bonding. This type of contact allows both sliding and gap between selected faces. Select the **Bonded** option if the selected surfaces are bonded together. In case of metals, you can assume that the two surfaces are welded together. Select the **Sliding/No Separation** option if you want the objects to be bonded at selected surfaces in such a way that they cannot move up or down but they can slide over one another. Note that friction is not applied in case of this contact. Select the **Separation/No Sliding** option if you want components to form gap in case of load application but do not want them to slide over one another. This option is applicable for non-linear analyses only. Select the **Offset Bonded** option if you want the objects to be welded at selected surfaces while maintaining specified gap between them. Select the **Shrink Fit/Sliding** option if you want to apply initial interference between the part and allow sliding. Select the **Shrink Fit/No Sliding** option if you want to apply initial interference between the part and do not allow sliding between parts. Select the **Disable** option from the **Type** drop-down to disable any contact between selected surfaces.
- Select the desired option from the **Penetration Type** drop-down to define penetration of master slave nodes during analysis. This drop-down is not available if **Edge to Surface** option is selected in **Type** drop-down or **Offset Bonded**, **Shrink Fit/Sliding**, or **Shrink Fit/No Sliding** option is selected in the **Contact Type** drop-down of the dialog box. There are two options available in **Penetration Type** drop-down viz. **Unsymmetric Contact** and **Symmetric Contact**. Select the **Unsymmetric Contact** option

if you want to check the penetration of slave nodes in master node. This option produces fast result at the loss of accuracy. Select the **Symmetric Contact** option if you want to check the penetration of both master and slave nodes. This option uses extra resources to produce result hence increasing the processing time.

- Specify the desired value of stiffness factor at contact site in the **Stiffness Factor** edit box.
- Specify the desired value of friction coefficient at the contact site in **Coefficient of Friction** edit box.
- Specify the desired value in **Penetration Surface Offset (mm)** edit box to define the depth upto which two surfaces will penetrate. Make sure to provide a realistic value otherwise the analysis will not converge.
- Set the parameters in other edit boxes as discussed earlier.
- Click in the **Master Entity** selection box and select the contacting face of first component.
- Click in the **Slave Entity** selection box and select the contacting face of second component; refer to Figure-61.

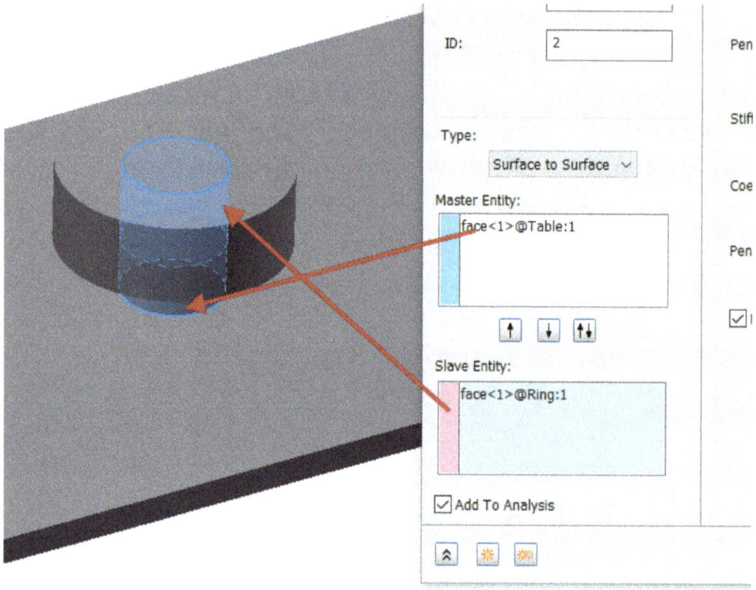

Figure-61. Faces selected for contact

- Click on the **Advanced Options** button to set advanced parameters for contact. The **Contact Parameters** dialog box will be displayed; refer to Figure-62.

Figure-62. Contact Parameters dialog box

- Specify desired values for various parameters and click on the **OK** button.

Setting Solver Options

The **Solver** tool in the **Contacts** panel is used to specify the region in which contacts will be applied automatically and the parameters like stiffness factor, friction coefficient, and so on. The procedure to use this tool is given next.

- Click on the **Solver** tool from the **Contacts** panel in the **Autodesk Inventor Nastran** tab of the **Ribbon**. The **Surface Contact** dialog box will be displayed as shown in Figure-63.

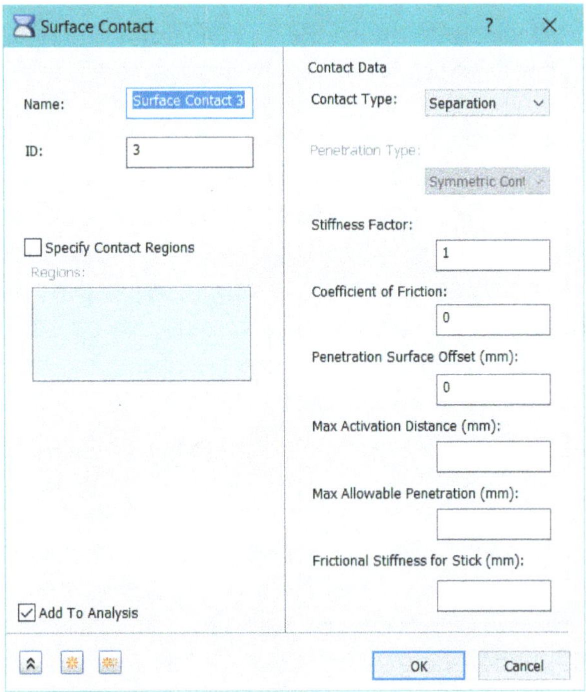

Figure-63. Surface Contact dialog box with Solver options

- Select the contact type you want to apply from the **Contact Type** drop-down.
- By default, complete model is analyzed to apply contacts. If you want to define a specific region in which contacts will be applied automatically then select the **Specify Contact Regions** check box. The **Regions** selection box will become active.
- Select the edges/faces/solids to define region; refer to Figure-64.

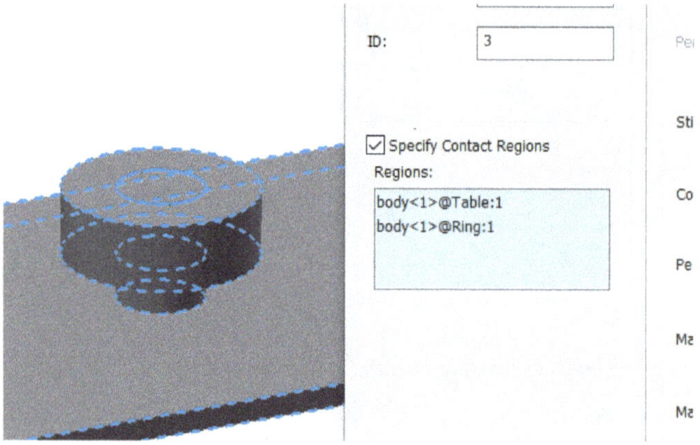

Figure-64. Region selected to define auto contacts

- After setting desired parameters, click on the **OK** button. The contact will be created.

MESHING

Meshing is the base of FEM. Meshing divides the solid/shell models into elements of finite size and shape. These elements are joined at some common points called nodes. These nodes define the load transfer from one element to other element. Meshing is a very crucial step in design analysis. The automatic mesher in the software generates a mesh based on specified global element size, tolerance, and local mesh control specifications. Mesh control lets you specify different sizes of elements for components, faces, edges, and vertices.

The software estimates a global element size for the model taking into consideration its volume, surface area, and other geometric details. The size of generated mesh (number of nodes and elements) depends on the geometry and dimensions of the model, element size, mesh tolerance, mesh control, and contact specifications. In the early stages of design analysis where approximate results may suffice, you can specify a larger element size for a faster solution. For a more accurate solution, a smaller element size may be required.

Meshing generates 3D tetrahedral solid elements, 2D triangular shell elements, and 1D beam elements. A mesh consists of one type of elements unless the mixed mesh type is specified. Solid elements are naturally suitable for bulky models. Shell elements are naturally suitable for modeling thin parts (sheet metals), and beams are suitable for modeling structural members. Autodesk Inventor Nastran uses linear tetrahedral (4 nodes) and parabolic tetrahedral (10 nodes) elements for solid objects. For Shell objects, triangular and quadrilateral elements are available with linear and parabolic orders. Linear triangular elements have 3 nodes, parabolic triangular elements have 6 nodes, linear quadrilateral elements have 4 nodes and parabolic quadrilateral elements have 8 nodes. For line objects, there are two types of elements available in Autodesk Inventor Nastran Beam and Bar.

Defining Mesh Settings

The **Mesh Settings** tool in **Mesh** panel is used to define global mesh size, element order, tolerance and other related parameters. The procedure to apply mesh settings is given next.

- Click on the **Mesh Settings** tool from the **Mesh** panel in **Autodesk Inventor Nastran** tab of **Ribbon**. The **Mesh Settings** dialog box will be displayed; refer to Figure-65.

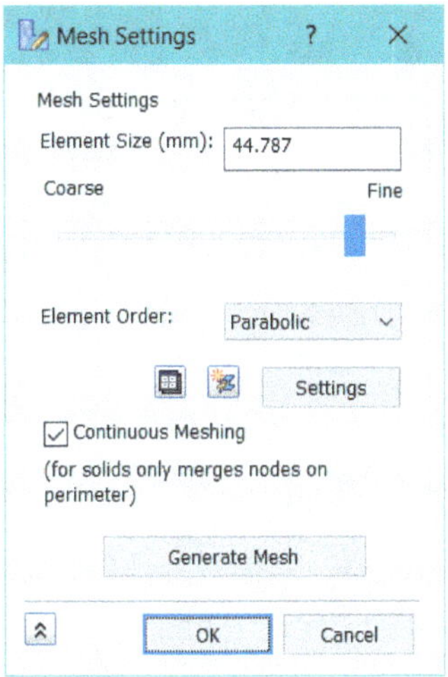

Figure-65. Mesh Settings dialog box

- Specify the desired value of element size in the **Element Size** edit box. You can also define mesh size by using the slider. Moving slider towards **Coarse** makes the mesh element size large and moving slider towards **Fine** makes the mesh element size smaller.
- Select the **Parabolic** option from the **Element Order** drop-down to create 2nd order elements. Select the **Linear** option from the **Element Order** drop-down to create 1st order elements. It is better to select **Parabolic** option for 3D objects and **Linear** option for 2D surfaces or linear objects.
- Click on the **Settings** button to define advanced parameters for meshing. The **Advanced Mesh Settings** dialog box will be displayed; refer to Figure-66.

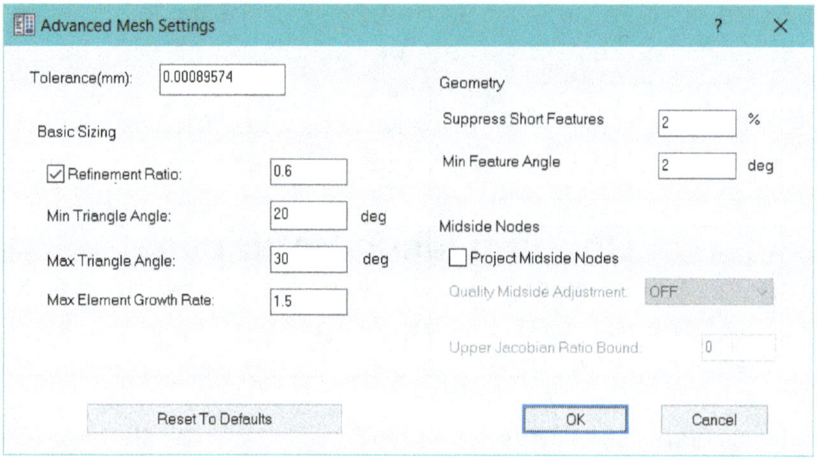

Figure-66. Advanced Mesh Settings dialog box

- Specify minimum tolerance value up to which mesh elements will be created.
- Select the **Refinement Ratio** check box to set mesh size more uniform. The larger the refinement ratio, the larger is the mesh size. Specify the desired refinement ratio in the **Refinement Ratio** edit box.
- Specify minimum triangle angle and maximum triangle angle values in **Min Triangle Angle** and **Max Triangle Angle** edit boxes respectively.
- Specify the desired element growth rate in **Max Element Growth Rate** edit box.
- Specify the desired value in **Suppress Short Features** edit box to define the percentage of small features like holes, bumps etc. to be suppressed for meshing.
- Specify the desired value in degree in **Min Feature Angle** edit box to specify the mesh angle around complex features like holes, edges etc. This is the minimum angle value to be considered for meshing below this value, the angle features of faces will be suppressed.
- Select the **Project Midside Nodes** check box to define parameters related to midside nodes. Midside nodes are created on elements along curves and edges of model. On selecting this check box, the options in the **Midside Nodes** area become active.
- Select the desired option from the **Quality Midside Adjustment** drop-down. Select **ON** option from the drop-down to automatically adjust mid side nodes for generating positive Jacobian ratio.
- Specify the desired value in **Upper Jacobian Ratio Bound** edit box to define the quality of midside nodes.
- After setting parameters, click on the **OK** button. The **Mesh Settings** dialog box will be displayed again.

Mesh Table Parameters

- Click on the **Mesh Table** button from the **Mesh Settings** dialog box to define size of mesh. The **Mesh Table** dialog box will be displayed; refer to Figure-67. You can display the same dialog box by clicking on **Table** button in **Mesh** panel.

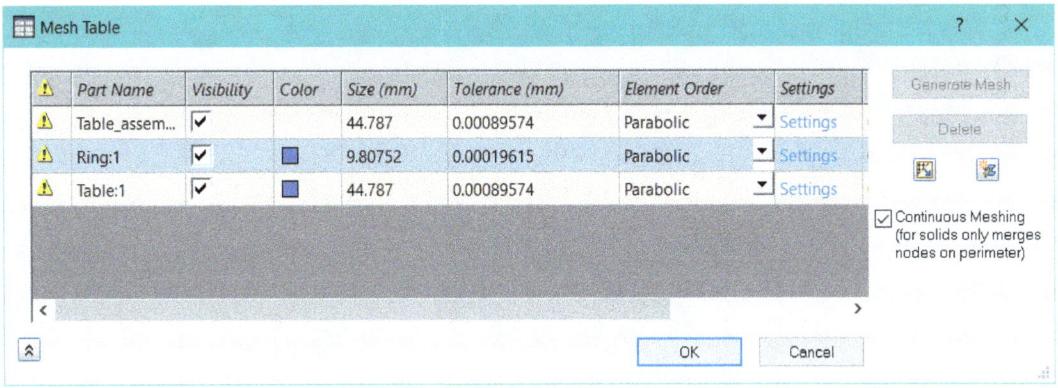

Figure-67. Mesh Table dialog box

- Click in the edit boxes under **Size** column and specify the size of mesh elements. Set the other parameters for each of the component in assembly in table and click on the **OK** button.

Generating Mesh

- After setting all the parameters, click on the **Generate Mesh** tool from the **Mesh** panel of **Autodesk Inventor Nastran** tab in **Ribbon**. An information box will be displayed and the mesh will be generated; refer to Figure-68.

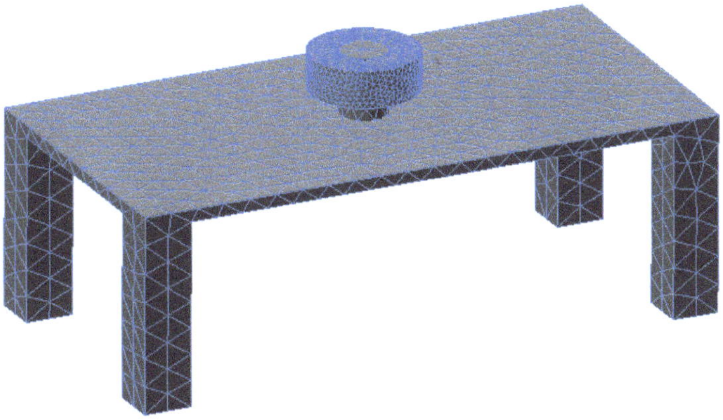

Figure-68. Mesh generated

Setting Mesh Control

The **Mesh Control** tool is used to set fine or coarse mesh at selected locations. The procedure to use this tool is given next.

- Click on the **Mesh Control** tool from the **Mesh** panel in **Autodesk Inventor Nastran** tab of **Ribbon**. The **Mesh Control** dialog box will be displayed; refer to Figure-69.

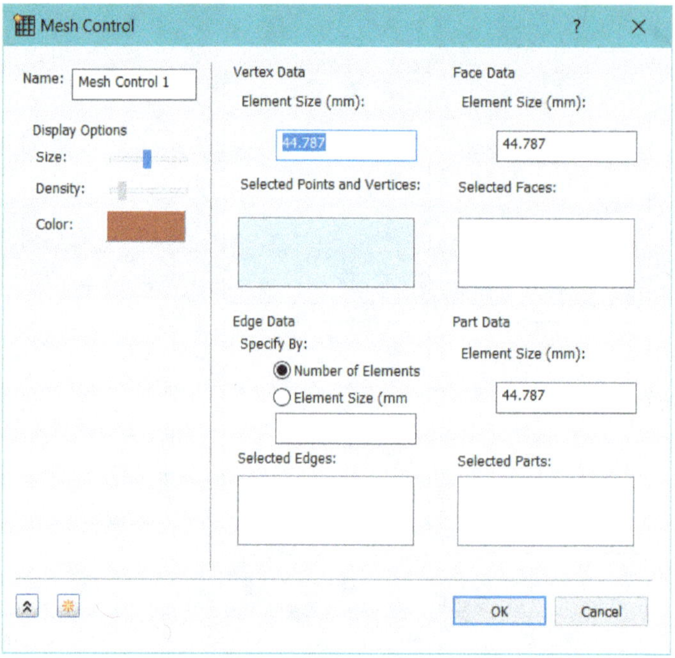

Figure-69. Mesh Control dialog box

- Click in the **Selected Points and Vertices** selection box and select the points/vertices on which you want to apply mesh control.
- Specify the desired element size in the **Element Size** edit box above the selection box to define mesh size.
- Similarly, you can use the other options to define mesh element sizes for faces, edges, or parts.
- After setting desired parameters, click on the **OK** button.

RUNNING ANALYSIS

Once you have set the parameters for analysis, click on the **Run** button from the **Solve** panel of **Autodesk Inventor Nastran** tab in the **Ribbon**. The system will start solving the analysis. Once the analysis is complete, an information box will be displayed; refer to Figure-70. After setting parameters, click on the **OK** button from the information box. The results will be displayed; refer to Figure-71.

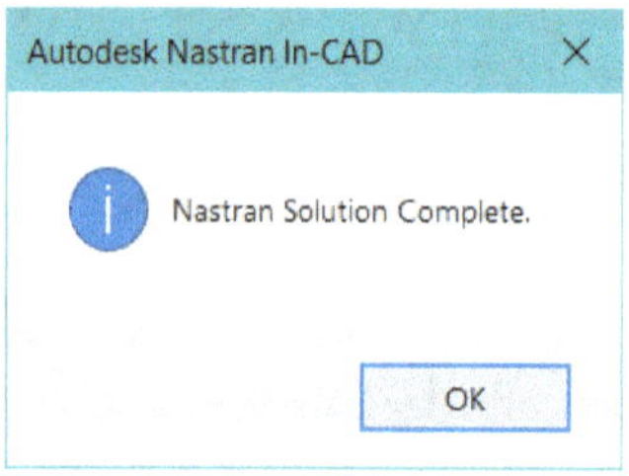

Figure-70. Autodesk Inventor Nastran information box

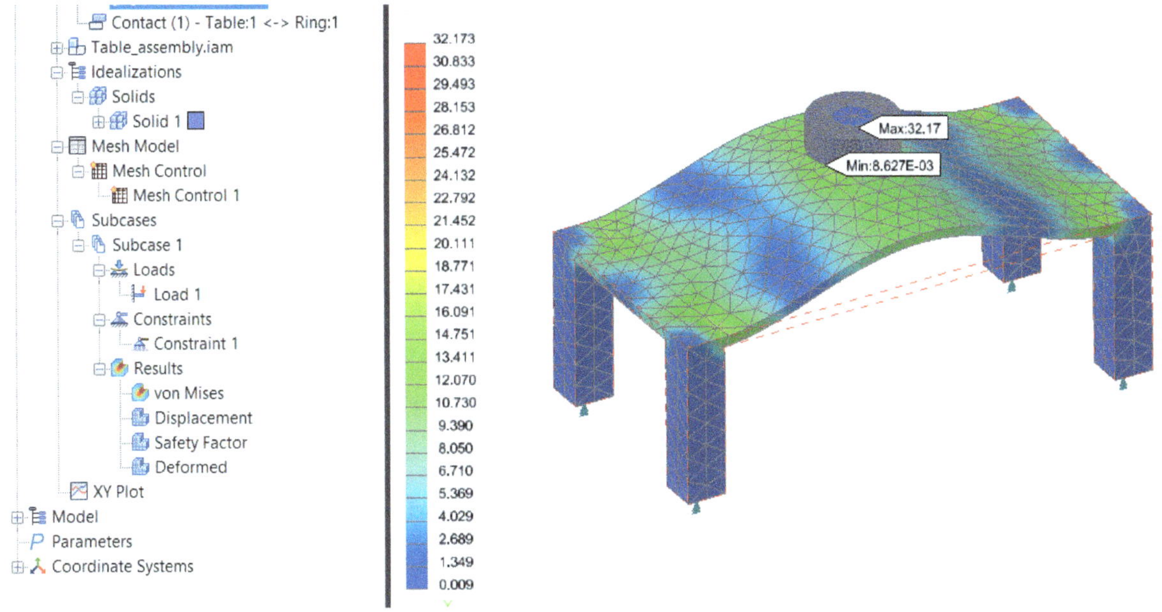

Figure-71. Result generated

To check any result, double-click on it from the **Results** node at the left of application window.

SETTING PLOT OPTIONS

The **Options** tool in the **Results** panel is used to set plot options. The procedure to set plot options is given next.

- Select the desired sub-case from the **Subcases** drop-down in the **Results** panel of **Autodesk Inventor Nastran** tab. Click on the **Options** tool from the **Results** panel of **Autodesk Inventor Nastran** tab in **Ribbon**. The **Plot** dialog box will be displayed; refer to Figure-72.

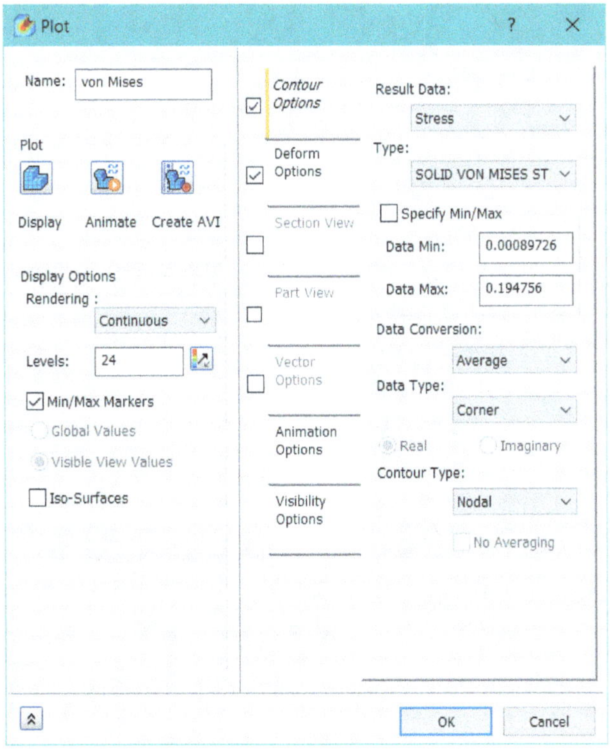

Figure-72. Plot dialog box

Contour Options

- Select the **Contour Options** check box to set contour options. The options related to contour will be displayed.
- Select the **Stress** option from the **Result Data** drop-down to create stress plot. Select the **Displacement** option if you want to create displacement plot. Similarly, use the **Strain**, **Reaction Force**, and **Contact** options to create strain, reaction force, and contact plots.
- If you want to set other plots like mesh convergence error, safety factor, and so on then select the **Other** option from the **Results Data** drop-down and select the respective options from the **Type** drop-down. You can select the desired type for stress, strain, and other results from the **Type** drop-down.
- Select the **Specify Min/Max** check box to specify minimum and maximum values of parameter for the plot in **Data Min** and **Data Max** edit boxes.
- Select the **Nodal** option from the **Contour Type** drop-down if you want to generate result node-wise. Select the **Elemental** option if you want to generate results element-wise. If **Elemental** option is selected then select the **No Averaging** check box to create a discrete plot of result.
- Specify the other parameters as required for contour options.

Deform Options

- Select the **Deform Options** check box to manage deformation of model in result. The options will be displayed as shown in Figure-73.

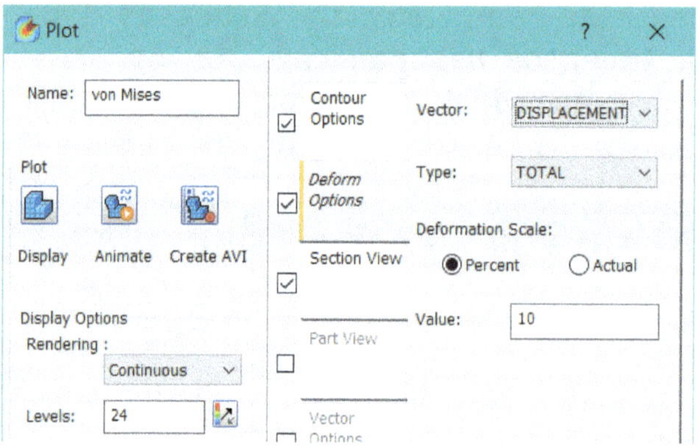

Figure-73. Deform Options

- Select the desired option from the **Vector** drop-down to define the parameters to taken into consideration for deformation in plot.
- Select the desired component of selected parameter from the **Type** drop-down.
- Select the **Percent** or **Actual** radio button as required and specify the desired value for deformation scale.

Section View Options

- Select the **Section View** check box to create section of model in result plots. The options will be displayed as shown in Figure-74.

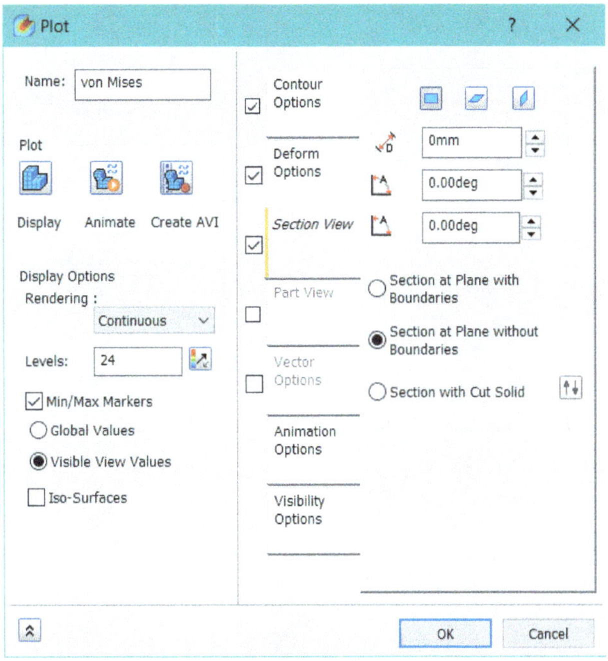

Figure-74. Section View options

- Select the desired button to define which plane is to be used for section.
- Specify the desired distance and angle for plane to create section view.
- Set the other parameters as required; refer to Figure-75.

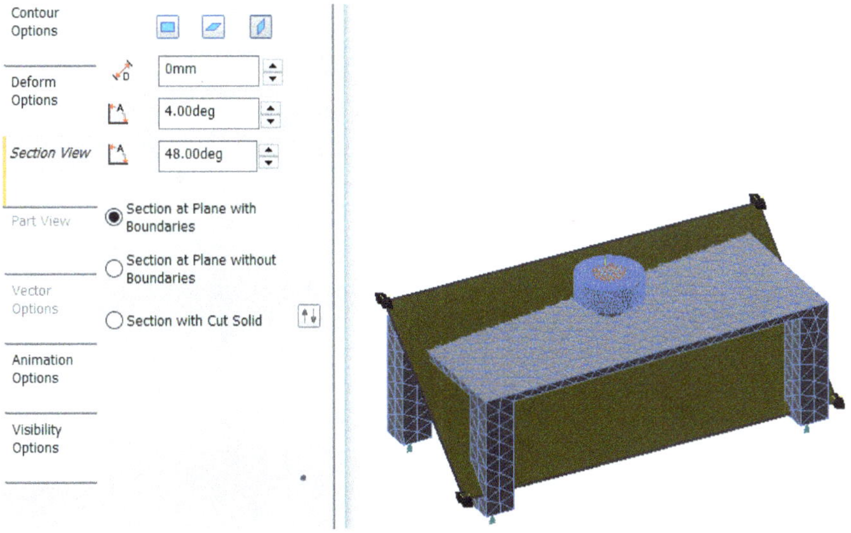

Figure-75. Section view parameters

Setting Part View Options

- Select the **Part View** check box to select the parts to be displayed in the result plots.
- The **Select Parts** selection box will be activated; refer to Figure-76.

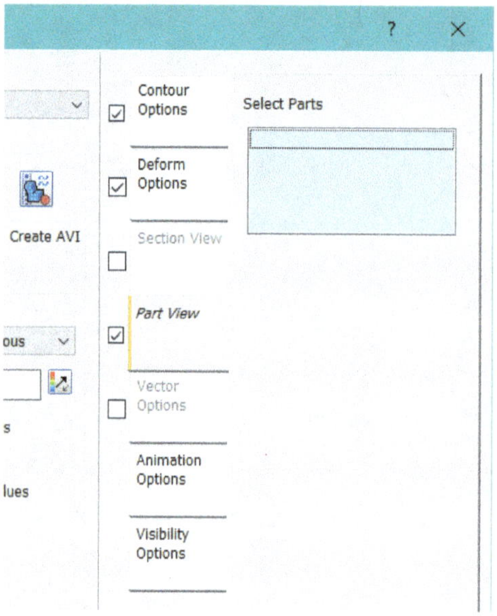

Figure-76. Part View options

Vector Options

The **Vector Options** are used to define the vector to be displayed in the result plot. The procedure is given next.

- Select the **Vector Options** check box from the **Plot** dialog box. The options will be displayed as shown in Figure-77.

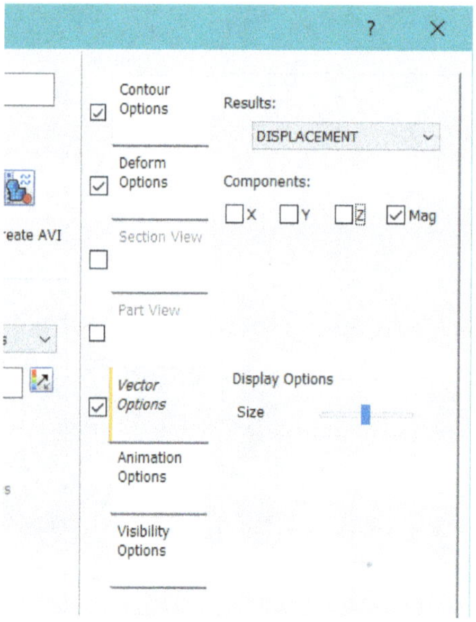

Figure-77. Vector Options

- Select the desired options to specify which vector is to be displayed in results.

Animation Options

- Select the **Animation Options** tab from the **Plot** dialog box. The options will be displayed as shown in Figure-78.

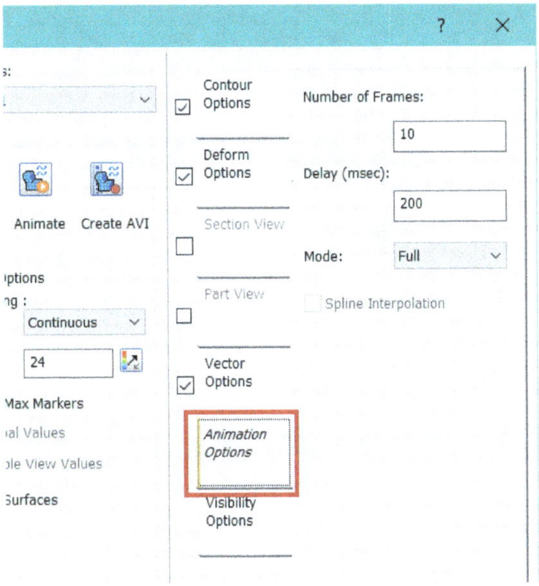

Figure-78. Animation Options

- Specify the number of frames in animation, delay between two frames of animation, and modes as required.

Visibility Options

- Click on the **Visibility Options** tab in the dialog box. The options will be displayed as shown in Figure-79.

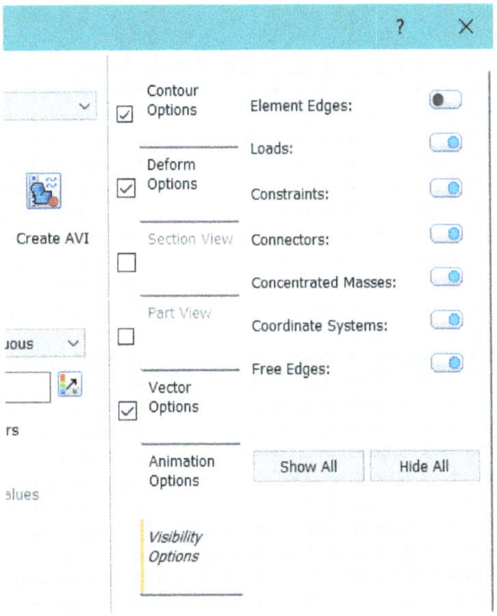

Figure-79. Visibility Options

- Select the desired radio buttons to set entities to be displayed in the result plot.

Now, there are three ways to generate plot viz. **Display**, **Animate**, and **Create AVI** buttons.

Click on the **Display** button to generate 2D image of result plot. Click on the **Animate** button to generate animation of result. The animation of result will run in the viewport. Click on the **Create AVI** button to save the animation. Once the recording is complete, an information box will be displayed; refer to Figure-80. Click on the **OK** button from the information box. The **Save** dialog box will be displayed; refer to Figure-81.

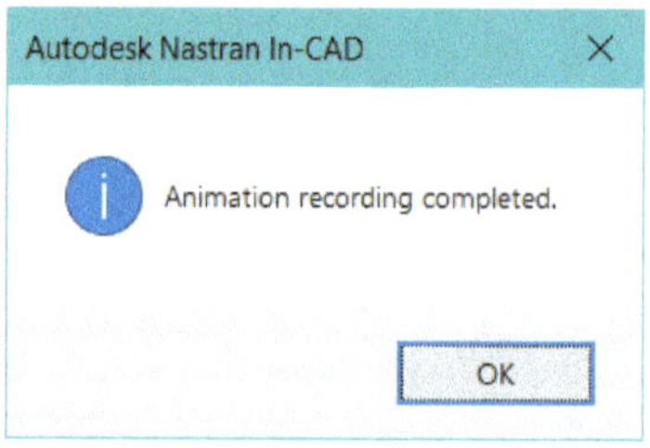

Figure-80. Information box

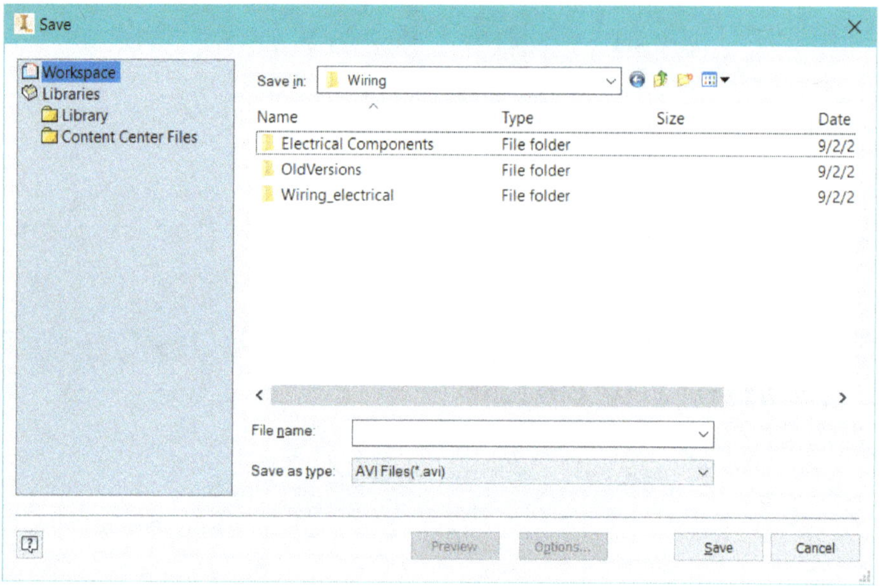

Figure-81. Save dialog box

Specify desired name of file in the **File name** edit box and click on the **Save** button. An information box will be displayed prompting you whether to open the file or not. Click on the **Yes** or **No** button as desired.

Using Probe to Check Node Results

Once the result is displayed, click on the **Probes** toggle button in **Results** panel of **Autodesk Inventor Nastran** tab in **Ribbon**. The cursor will be displayed with a plus sign. Hover the cursor at desired location on result model. The plot result will be displayed on respective node or element; refer to Figure-82.

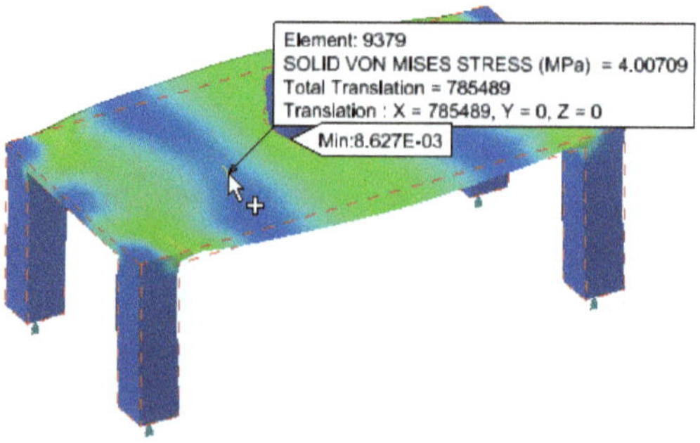

Figure-82. Using probe for result on node

Once the analysis is performed, click on the **Finish Autodesk Inventor Nastran** button from the **Exit** panel to exit analysis.

FOR STUDENT NOTES

FOR STUDENT NOTES

Chapter 3

Static Analyses

Topics Covered

The major topics covered in this chapter are:

- ***Introduction***
- ***Starting Linear Static Analysis***
- ***Applying Materials, Loads, and Constraints***
- ***Applying Idealizations***
- ***Setting Mesh and other parameters***
- ***Running Analysis and Plotting Results***
- ***Performing Linear Static Analysis on Assembly***
- ***Performing Prestress Static Analysis***
- ***Performing Nonlinear Static Analysis***

INTRODUCTION

The Linear Static analysis is performed to check the effect of static loads on the model. The process to create linear static study is given next.

PERFORMING LINEAR STATIC ANALYSIS ON PART

Perform linear static analysis on a steel bar of 22x10x120 mm with 1000 N load applied on its top flat face.

The procedure to perform linear static analysis is given next.

- Open the part on which you want to perform analysis in Autodesk Inventor; refer to Figure-1.

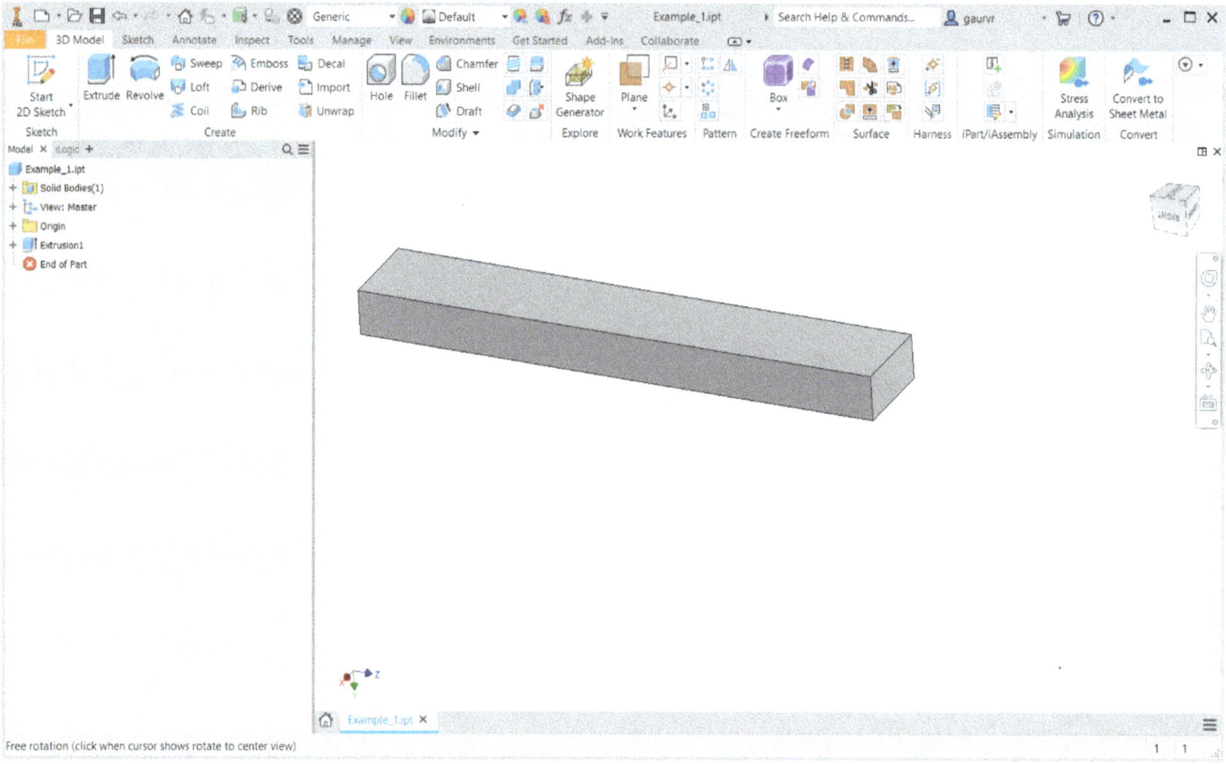

Figure-1. Example 1

- Click on the **Autodesk Inventor Nastran** tool from the **Environments** tab of **Ribbon**. The **Autodesk Inventor Nastran** tab will be added in the **Ribbon** and linear static analysis will start automatically.
- Click on the **Edit** button from the **Analysis** panel in **Autodesk Inventor Nastran** tab of **Ribbon**. The **Analysis** dialog box will be displayed; refer to Figure-2.

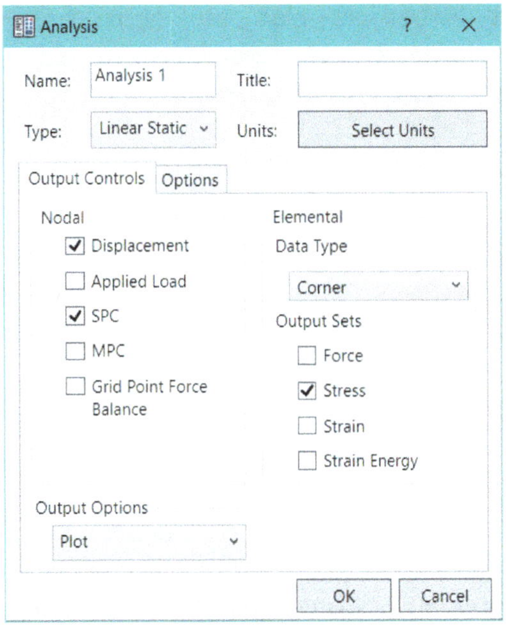

Figure-2. Analysis dialog box for linear static analysis

- Click on the **Select Units** button from the **Analysis** dialog box. The **Units** dialog box will be displayed.
- Select **SI** option from the **Unit System** drop-down in the dialog box and click on the **OK** button from **Units** dialog box.
- Click on the **OK** button from the **Analysis** dialog box.

Applying Material

- Click on the **Materials** tool from **Prepare** panel of **Autodesk Inventor Nastran** tab in the **Ribbon**. The **Material** dialog box will be displayed; refer to Figure-3.

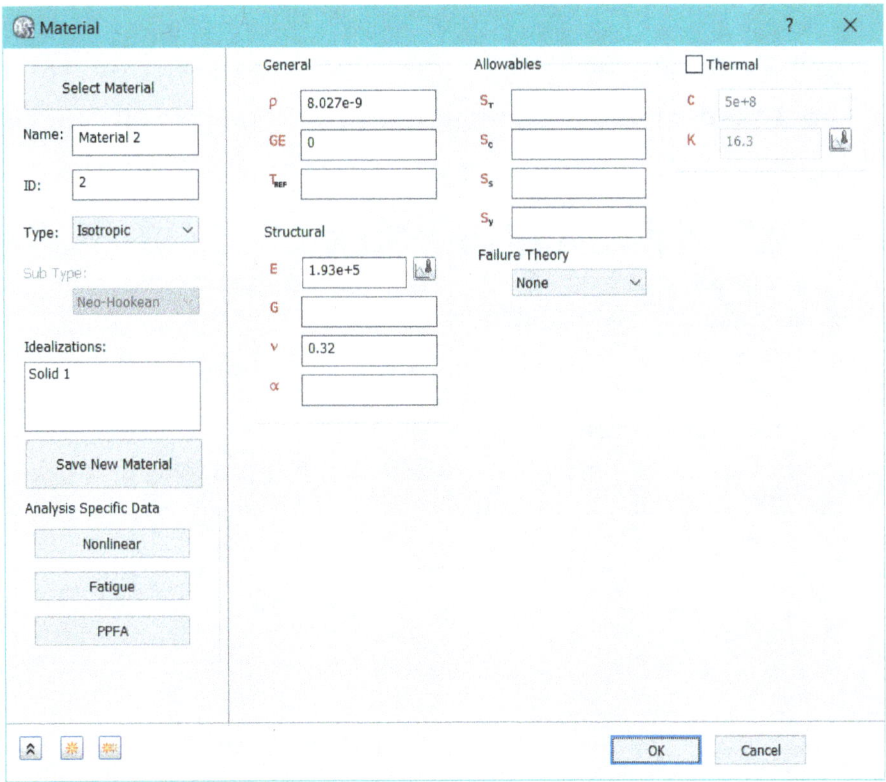

Figure-3. Material dialog box

- Click on the **Select Material** button from the **Material** dialog box. The **Material DB** dialog box will be displayed.
- Select the **Stainless Steel - Austenitic** material from the **Autodesk Material Library** node and click on the **OK** button.
- Click on the **OK** button from the **Material** dialog box to apply material.

Applying Idealizations

- Click on the **Idealizations** tool from the **Prepare** panel in **Autodesk Inventor Nastran** tab of **Ribbon**. The **Idealizations** dialog box will be displayed.
- Select **Stainless Steel-Austenitic** option from the **Material** drop-down and click on the **OK** button.

Applying Constraints

- Click on the **Constraints** tool from the **Setup** panel in **Autodesk Inventor Nastran** tab of **Ribbon**. The **Constraint** dialog box will be displayed.
- Select the faces as shown Figure-4.

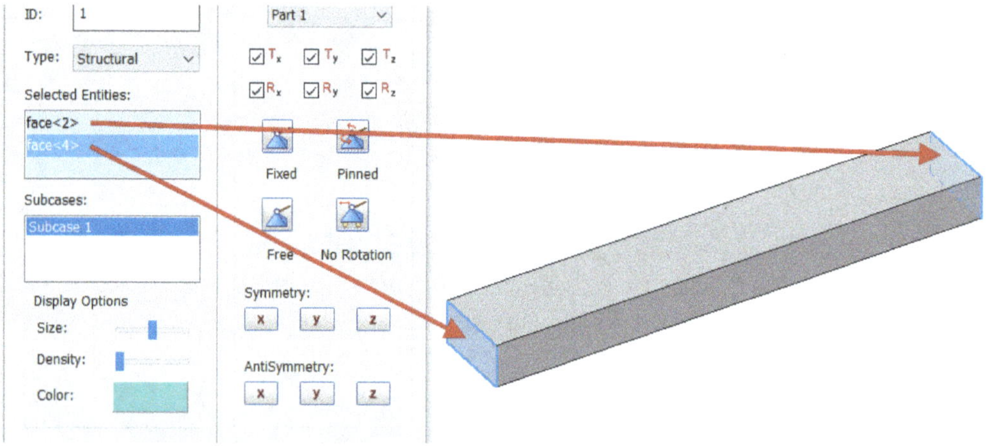

Figure-4. Faces selected for constraints

- Click on the **Fixed** button and then click on the **OK** button from the **Constraint** dialog box. The constraint will be applied.

Applying Force

- Click on the **Loads** button from the **Setup** panel of the **Ribbon**. The **Load** dialog box will be displayed.
- Select the top face of model to apply force; refer to Figure-5.
- Select **Force** option from the **Type** drop-down and **Normal To Surface** option from the **Direction** drop-down.
- Specify the value of force as **-1000** in **Magnitude** edit box.
- Click on the **OK** button from the dialog box to apply force.

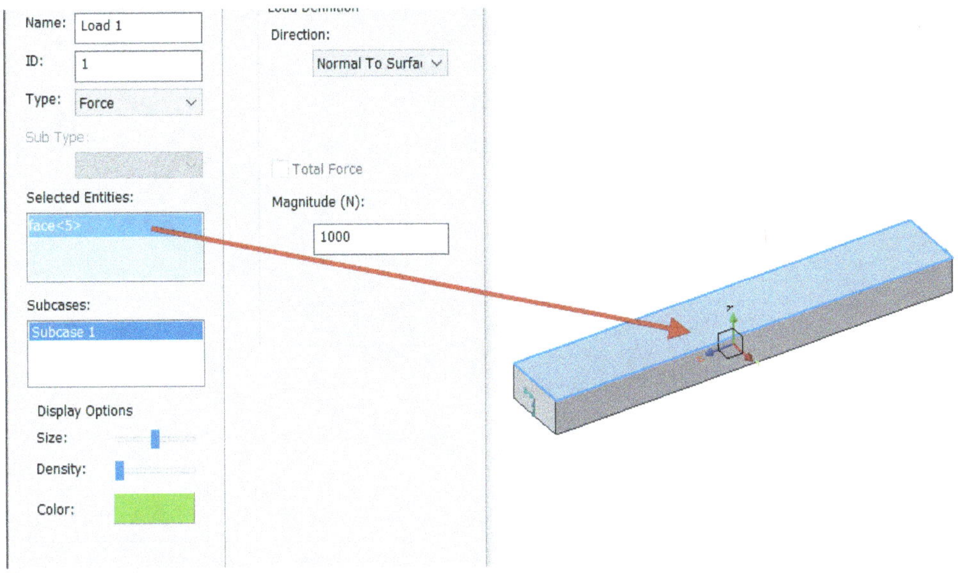

Figure-5. Applying force

Creating Mesh

- Click on the **Mesh Settings** tool from the **Mesh** panel of **Autodesk Inventor Nastran** tab in **Ribbon**. The **Mesh Settings** dialog box will be displayed.
- Specify the element size as **0.01** in **Element Size** edit box and click on the **Generate Mesh** button. The mesh will be created; refer to Figure-6.
- Click on the **OK** button to exit the dialog box.

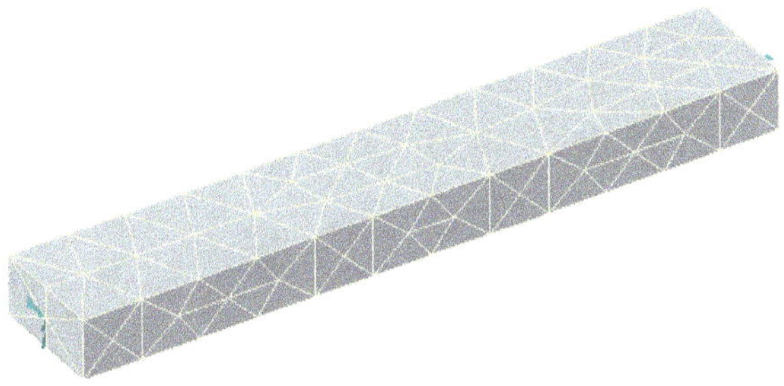

Figure-6. Mesh created

Running Analysis

- Click on the **Run** tool from the **Solve** panel in **Autodesk Inventor Nastran** tab of **Ribbon**. The system will start performing analysis and once complete an information box will be displayed.
- Click on the **OK** button from the information box. The analysis result will be displayed; refer to Figure-7.

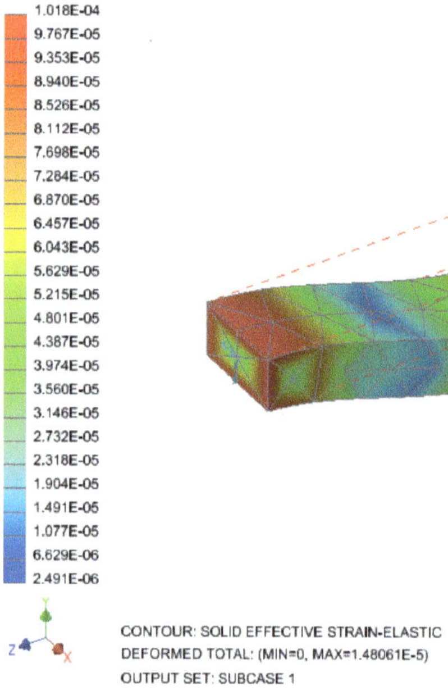

Figure-7. Analysis result created

- Click on the **Finish Autodesk Inventor Nastran** button from the **Exit** panel in **Autodesk Inventor Nastran** tab of **Ribbon** to exit the environment.

PERFORMING LINEAR STATIC ANALYSIS ON AN ASSEMBLY

Perform linear static study on the model shown in Figure-8. Apply bearing load at the connecting surface of table and rotor as per the figure.

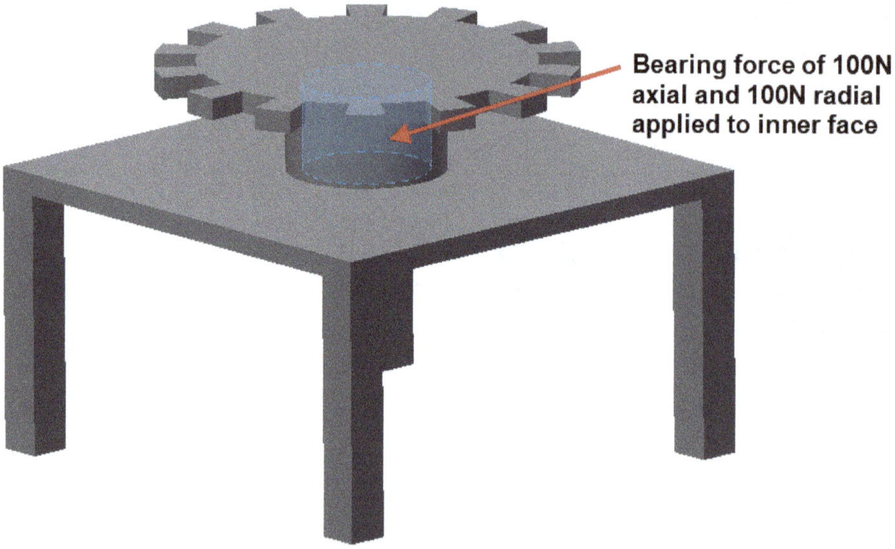

Figure-8. Assembly model for linear static analysis

Starting Linear Static Analysis

- Open the part in Autodesk Inventor and start Autodesk Inventor Nastran; refer to Figure-9.

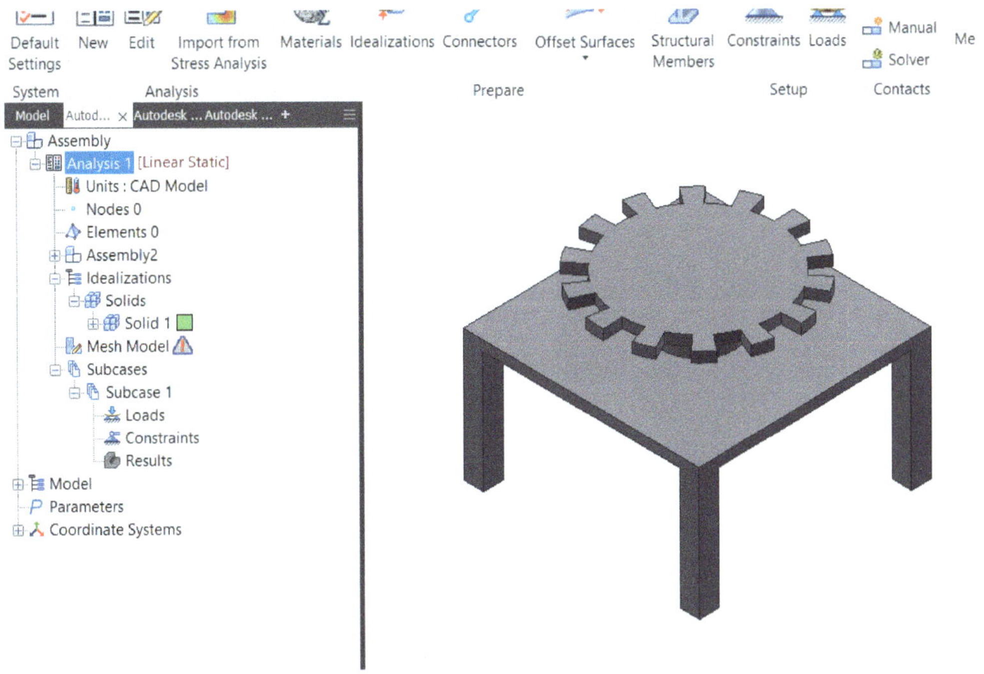

Figure-9. Assembly for linear static analysis

- Click on the **Edit** button from the **Analysis** panel of **Autodesk Inventor Nastran** tab in **Ribbon**. The **Analysis** dialog box will be displayed.
- In this analysis, we want the rotor to rotate under applied torque but do not want it to slide. So, we will use **Separation/No Sliding** as default contact. Click on the **Options** tab and select the **Separation/ No Sliding** option from the **Contact Type** drop-down. Select the **SI** units as discussed earlier.
- After setting the parameters, click on the **OK** button.

Applying Material

- Click on the **Materials** tool from the **Prepare** panel of **Autodesk Inventor Nastran** tab in the **Ribbon**. The **Material** dialog box will be displayed.
- Click on the **Select Material** button from the dialog box. The **Material DB** dialog box will be displayed.
- Select the **Steel AISI 1015** material from **Autodesk Material Library** and click on the **OK** button.
- Click on the **OK** button from the **Material** dialog box to apply material.
- Right-click on **Solid1** option from the **Idealizations** node and select the **Edit** option; refer to Figure-10. The **Idealizations** dialog box will be displayed. Select **Steel AISI 1015** option from the **Material** drop-down and click on the **OK** button.

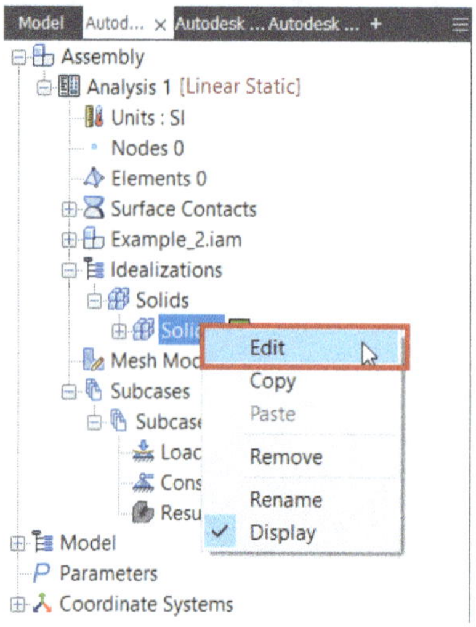

Figure-10. Editing idealization

Applying Contacts

- Click on the **Auto** tool from the **Contacts** panel in the **Ribbon**. The default contacts will be automatically applied in the assembly. Note that we have set **Separation/No Sliding** as default contact earlier.

Applying Constraint

- Click on **Constraints** tool from the **Setup** panel of **Autodesk Inventor Nastran** tab in the **Ribbon**. The **Constraint** dialog box will be displayed.
- Select the bottom faces of the table to apply fix constraint and click on the **Fixed** button; refer to Figure-11.

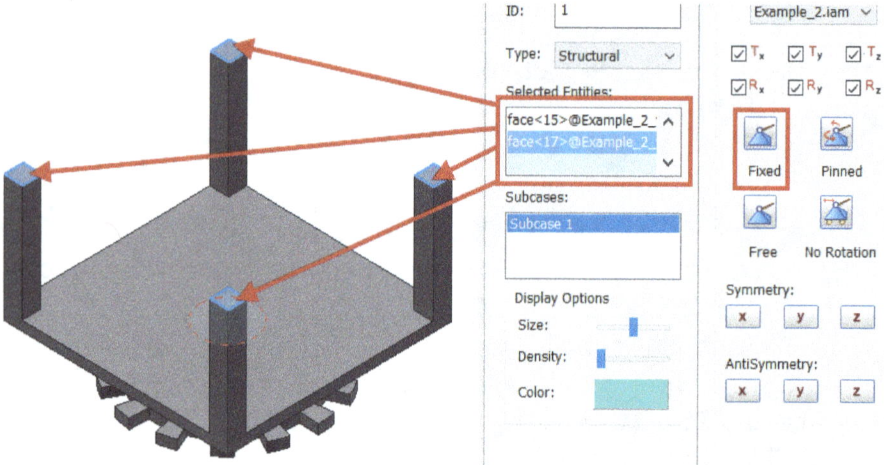

Figure-11. Faces selected for fixing

- Click on the **OK** button from the dialog box after applying constraint.

Applying Loads

- Click on the **Loads** tool from the **Setup** panel of **Autodesk Inventor Nastran** tab in **Ribbon**. The **Load** dialog box will be displayed.
- Select the round face of rotor as shown in Figure-12.

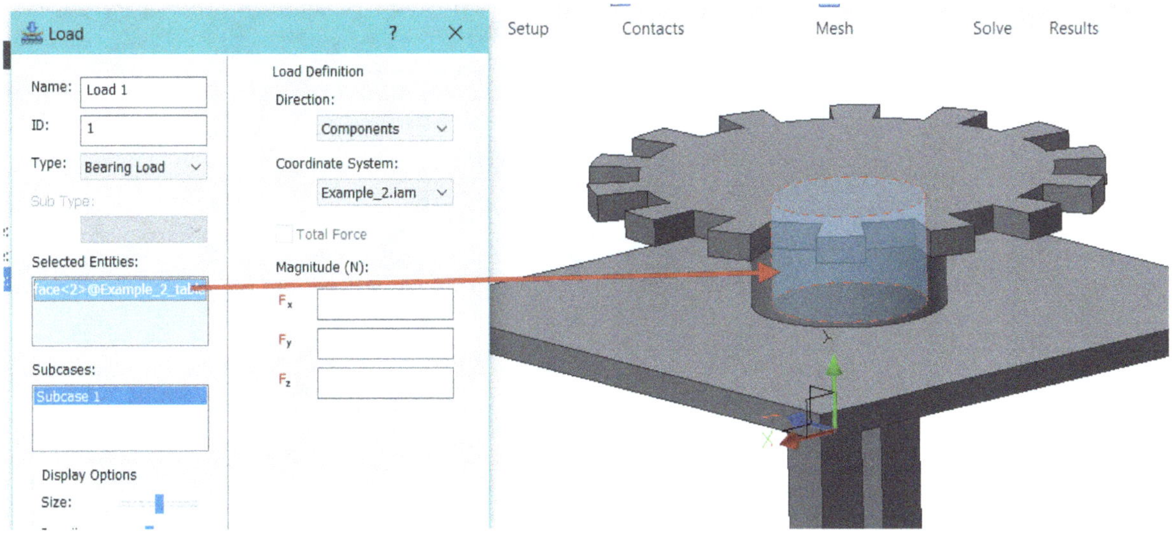

Figure-12. Applying bearing load

- Specify the radial and axial forces of bearing load as **500** in **F**$_x$ and **F**$_y$ edit boxes.
- Click on the **OK** button from the dialog box to apply load.

Mesh Setting

- Click on **Mesh Settings** tool from the **Mesh** panel of **Autodesk Inventor Nastran** tab in **Ribbon**. The **Mesh Settings** dialog box will be displayed.
- Specify the element size as **0.01** in **Element Size** edit box.
- Click on the **Generate Mesh** button from the dialog box. The mesh will be generated with specified mesh element size; refer to Figure-13.

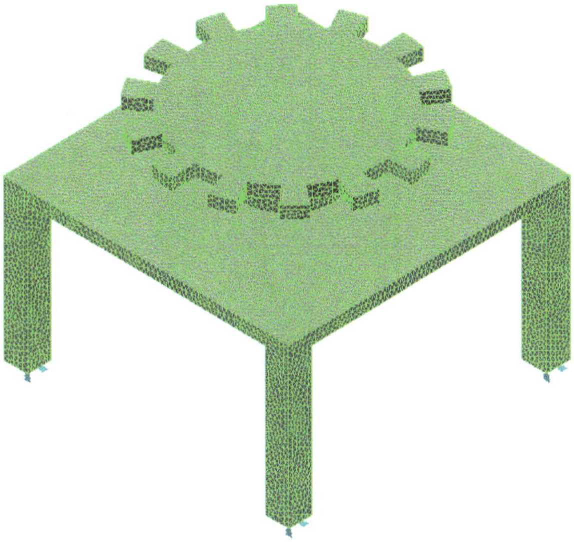

Figure-13. Mesh generated

- Click on the **OK** button from the dialog box.

Running Analysis

- Click on the **Run** button from the **Solve** panel of **Autodesk Inventor Nastran** tab in **Ribbon**. Once the analysis is complete, **Autodesk Inventor Nastran** information box will be displayed. Click on the **OK** button. The analysis results will be displayed; refer to Figure-14.

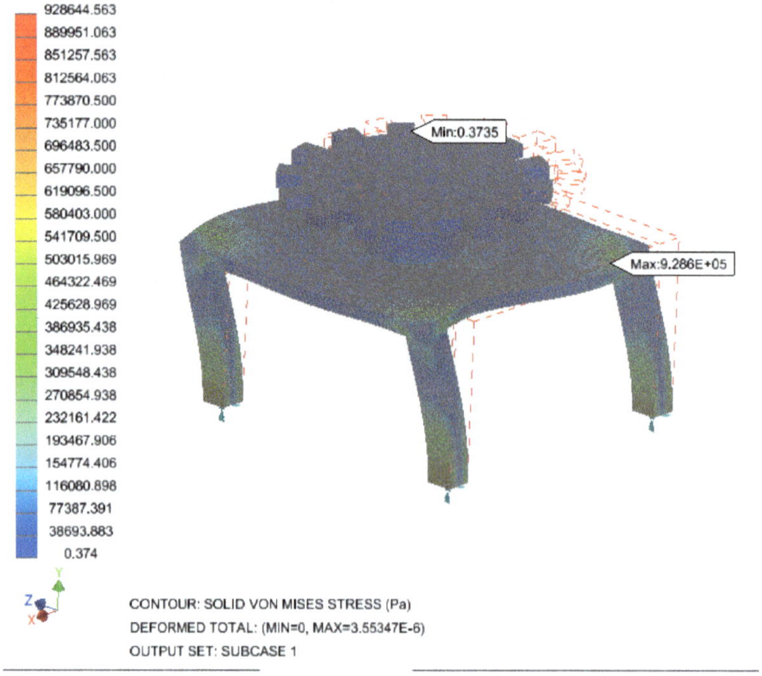

Figure-14. Analysis result of bearing load

- Double-click on desired option from the **Results** node.

PERFORMING ANALYSIS ON FRAME MODEL

When you open a frame model in Autodesk Inventor and start Autodesk Inventor Nastran then model features are automatically identified and beam idealization is applied to them; refer to Figure-15.

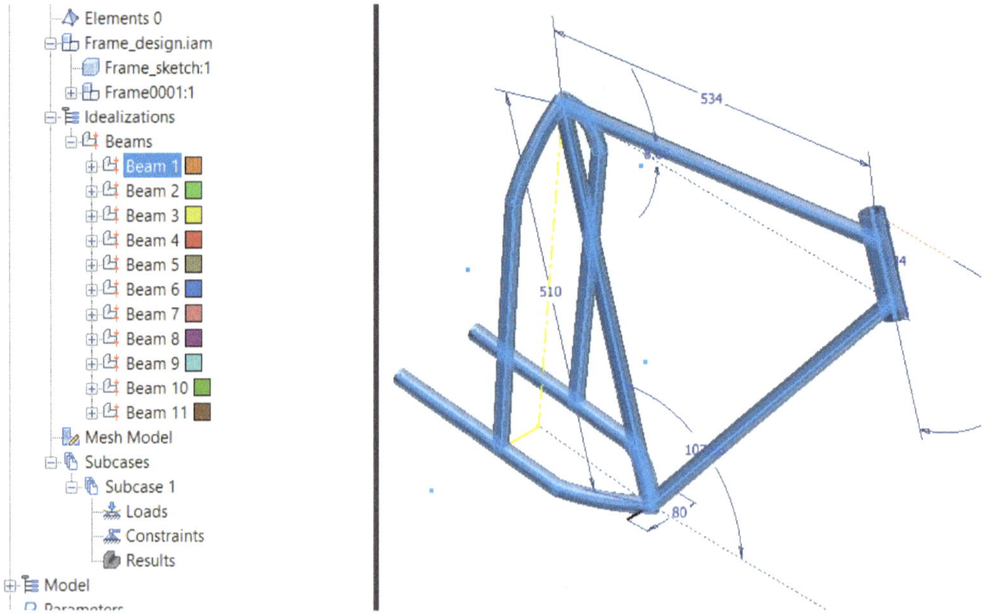

Figure-15. Beam idealization applied automatically

PERFORMING PRESTRESS STATIC ANALYSIS

Prestress static analysis is used when you are concerned about heavy pre-loading on the model. In case of prestress static analysis, two cases of same load are applied. The first case is used as pre-stressing load and the other is extra loading. The procedure to use this analysis is given next.

- Open the desired part and start Autodesk Inventor Nastran. By default, linear static analysis will be created.
- Click on the **Edit** button from the **Analysis** panel of **Autodesk Inventor Nastran** tab in the **Ribbon**. The **Analysis** dialog box will be displayed.
- Select the **Prestress Static** option from the **Analysis** drop-down in the dialog box. Set the other options as required and click on the **OK** button. Parameters for study will be displayed with two load sub-cases automatically created; refer to Figure-16.

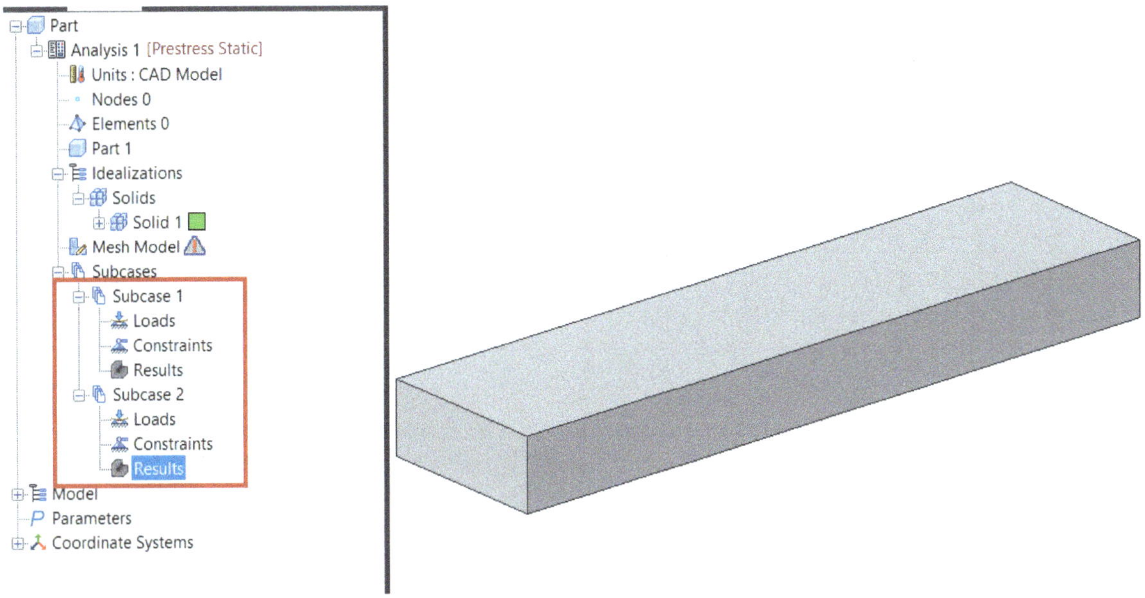

Figure-16. Prestress static analysis parameters

- Right-click on the **Loads** option for **Subcase 1** and select the **New** option from the shortcut menu. The **Load** dialog box will be displayed.
- Select the face on which you want to apply load and specify the desired value of load.
- Click on the **OK** button. The specified load will be applied as initial condition.
- Create additional loads as required in Subcase 2.
- Set the other parameters as required and run the analysis.

NON LINEAR STATIC ANALYSIS

All real structures behave nonlinearly in one way or another at some level of loading. In some cases, linear analysis may be adequate. In many other cases, the linear solution can produce erroneous results because the assumptions upon which it is based are violated. Nonlinearity can be caused by the material behavior, large displacements, and contact conditions.

You can use a nonlinear study to solve a linear problem. The results can be slightly different due to different procedures.

In nonlinear finite element analysis, a major source of nonlinearity is due to the effect of large displacements on the overall geometric configuration of structures. Structures undergoing large displacements can have significant changes in their geometry due to load-induced deformations which can cause the structure to respond nonlinearly in a stiffening and/or a softening manner.

For example, cable-like structures generally display a stiffening behavior on increasing the applied loads while arches may first experience softening followed by stiffening, a behavior widely-known as the snap-through buckling.

Another important source of nonlinearity stems from the nonlinear relationship between the stress and strain which has been recognized in several structural behaviors. Several factors can cause the material behavior to be nonlinear. The dependency of the material stress-strain relation on the load history (as in plasticity problems), load duration (as in creep analysis), and temperature (as in thermoplasticity) are some of these factors.

This class of nonlinearity, known as material nonlinearity, can be idealized to simulate such effects which are pertinent to different applications through the use of constitutive relations.

A special class of nonlinear problems is concerned with the changing nature of the boundary conditions of the structures involved in the analysis during motion. This situation is encountered in the analysis of contact problems.

Pounding of structures, gear-tooth contacts, fitting problems, threaded connections, and impact bodies are several examples requiring the evaluation of the contact boundaries. The evaluation of contact boundaries (nodes, lines, or surfaces) can be achieved by using gap (contact) elements between nodes on the adjacent boundaries.

Yielding of beam-column connections during earthquakes is one of the applications in which material nonlinearity are plausible.

THEORY BEHIND NON-LINEAR ANALYSIS

As you have learnt from introduction, there are three factors which cause non-linearity in the body which are elastic behavior of material, large displacement in body, and variation in contact. In terms of equation this can be expressed as

$$[K(D)]\{D\} = \{F\}$$

Here, K(D) is stiffness matrix which is function of displacement.

D is the displacement matrix
and F is the Force applied

So, from the above equation you can easily understand that the stiffness is changing according to the displacement. Note that displacement is vector and it also has a direction. If we compare the linear static analysis and non-linear static analysis curves then we can find out different effects of load in both conditions; refer to Figure-17.

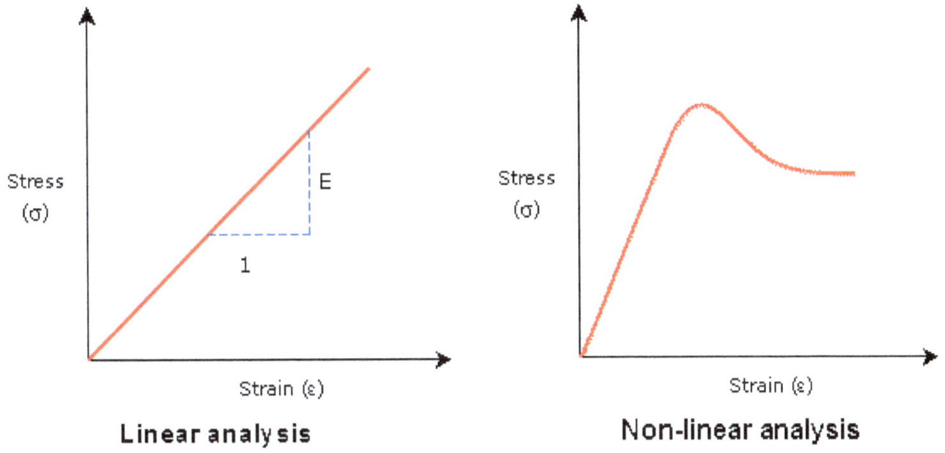

Figure-17. Stress Strain diagrams

In terms of load and displacement, the curve for both analysis can be given by Figure-18.

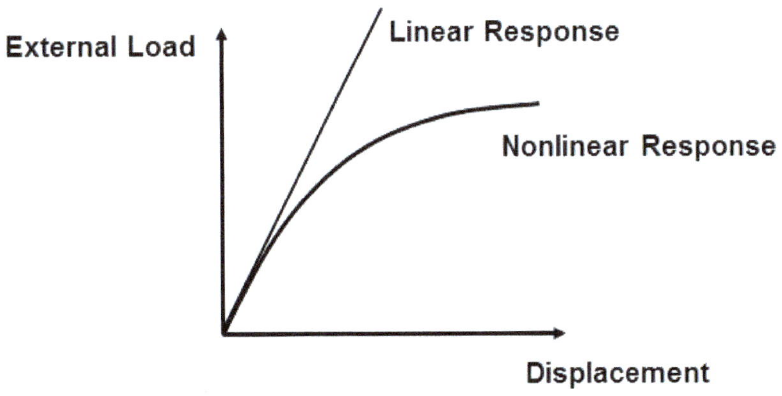

Figure-18. Load displacement curve

The normal incremental iteration does not work good for the non-linear analysis and generate errors as shown in Figure-19. So, Newton-Raphson algorithm is used to solve the non-linear equation.

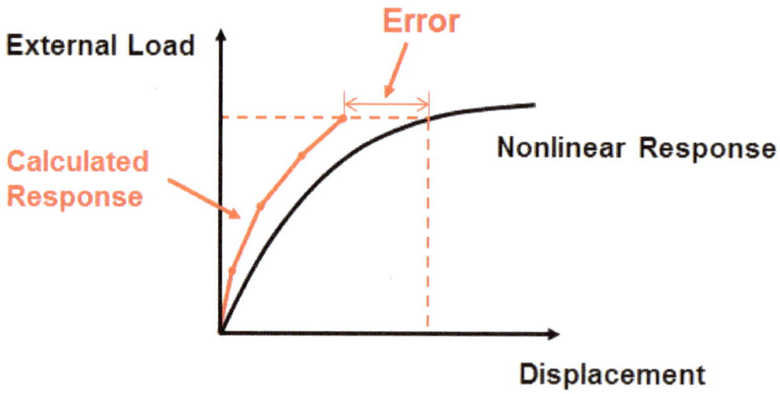

Figure-19. Error in incremental interative method

The equation for Newton-Raphson method is given as:

$$[K_T]\{\Delta u\} = \{F\} - \{F^{nr}\}$$

$[K_T]$ = tangent stiffness matrix
$\{\Delta u\}$ = displacement increment
$\{F\}$ = external load vector
$\{F^{nr}\}$ = internal force vector

The iteration continues till {F} - {Fnr} (difference between external and internal loads) is within a tolerance; refer to Figure-20.

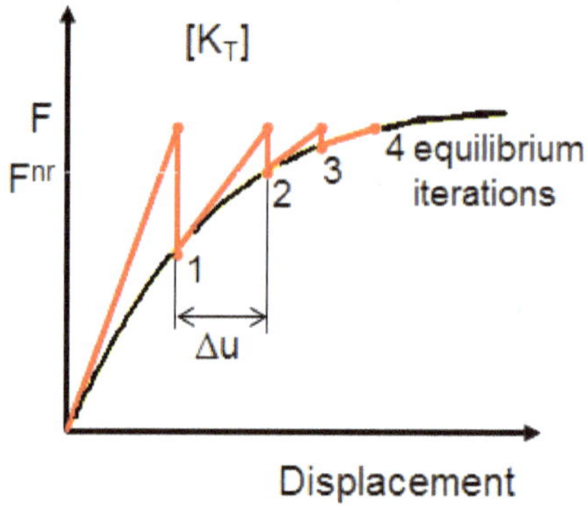

Figure-20. Newton-Raphson method

Thus a nonlinear solution typically involves the following:

- One or more load steps to apply the external loads and boundary conditions.(This is true for linear analyses too.)
- Multiple sub-steps to apply the load gradually. Each sub-step represents one load increment.(A linear analysis needs just one sub-step per load step.)
- Equilibrium iterations to obtain equilibrium (or convergence) at each sub-step.(Does not apply to linear analyses.)

Role of Time in Non-Linear Analysis

Each load step and sub-step is associated with a value of time. Time in most nonlinear static analyses is simply used as a counter and does not mean actual, chronological time. Figure-21 shows a load-time curve.

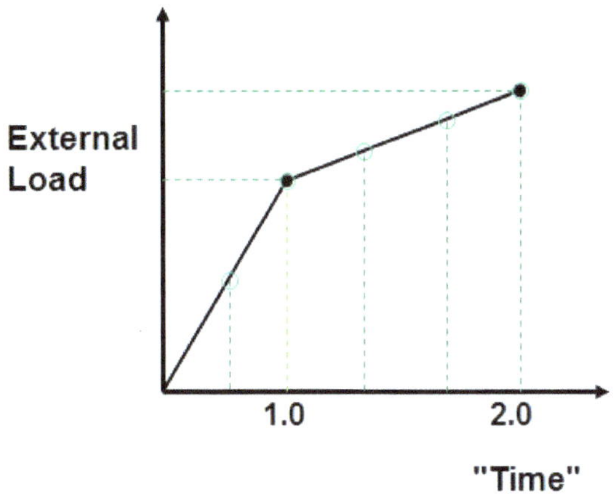

Figure-21. Load time curve

By default, time = 1.0 at the end of load step 1, 2.0 at the end of load step 2, and so on. For rate-independent analyses, you can set it to any desired value for convenience. For example, by setting time equal to the load magnitude, you can easily plot the load-deflection curve.

The "time increment" between each sub-step is the time step Δt. Time step Δt determines the load increment ΔF over a sub-step. The higher the value of Δt, the larger the ΔF, so Δt has a direct effect on the accuracy of the solution.

Now, we are going to start with Non-linear Static Analysis which means there is no damping or resistance to the force. Note that the last example in previous chapter, an assembly of wall bracket, was a problem of Nonlinear static analysis because there was large displacement in the body. Now, we will learn the procedure to perform Nonlinear static analysis.

PERFORMING NON LINEAR STATIC ANALYSIS

Basic steps to perform a nonlinear static analysis are given next.

- Open the part on which you want to perform nonlinear static analysis in Autodesk Inventor.
- Click on the **Autodesk Inventor Nastran** tool from the **Environments** tab in the **Ribbon**. The **Autodesk Inventor Nastran** tab will be added in the **Ribbon** and tools related to analysis will be displayed. Note that linear static analysis starts automatically when you activate the Nastran environment.
- Click on the **Edit** tool from the **Analysis** panel in **Autodesk Inventor Nastran** tab of **Ribbon**. The **Analysis** dialog box will be displayed.

- Select the **Nonlinear Static** option from the **Type** drop-down in the dialog box.
- Set the desired options in **Output Controls** tab of the dialog box.
- Click on the **Options** tab in the dialog box and select the **On** option from the **Large Displacements** drop-down if large displacement is expected in analysis; refer to Figure-22.

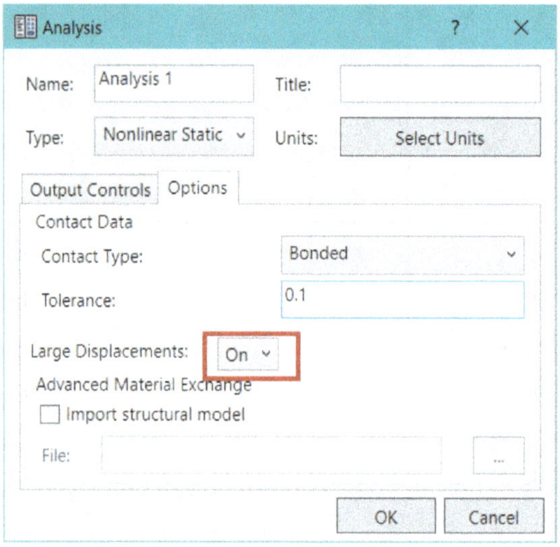

Figure-22. Large Displacements option

- If you want to import structural model from a nastran file earlier created then select the **Import structural model** check box and open the desired file using browse button.
- After setting the desired parameters, click on the **OK** button from the dialog box. The options in the analysis browser will be displayed as shown in Figure-23.

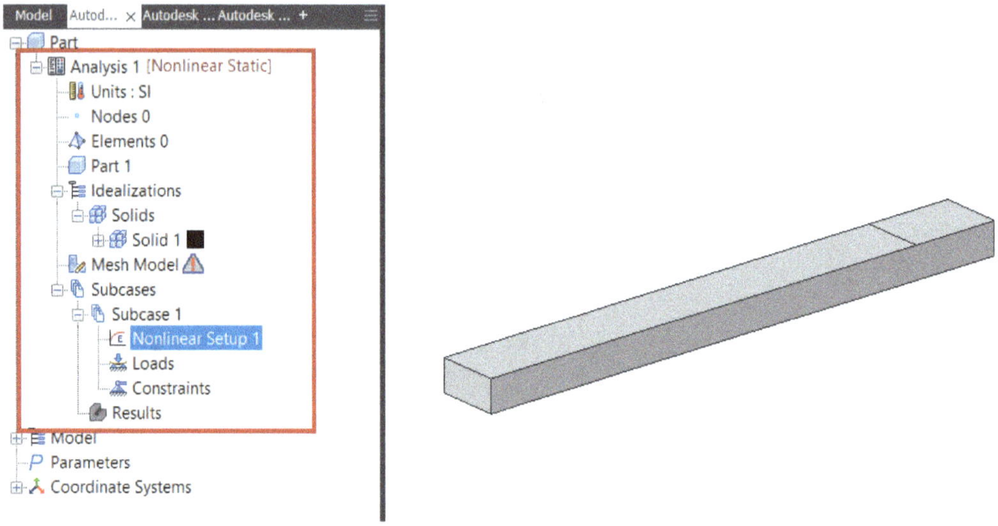

Figure-23. Options for nonlinear analysis

Nonlinear Setup Parameters

- Right-click on **Nonlinear Setup 1** option from the **Subcase 1** node and select the **Edit** option. The **Nonlinear Setup** dialog box will be displayed; refer to Figure-24.

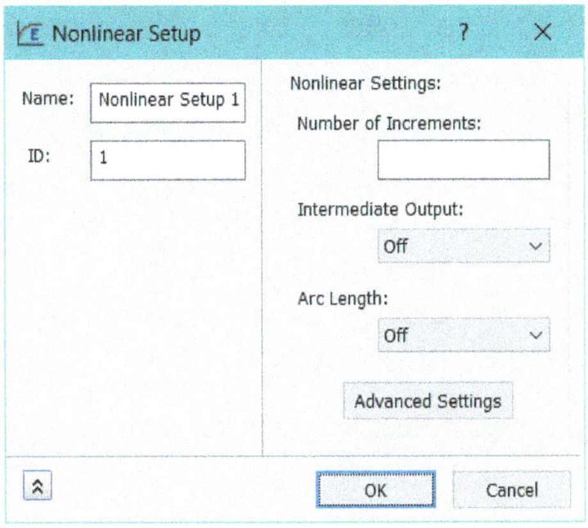

Figure-24. Nonlinear Setup dialog box

- Specify the desired number of steps in which load will be increased during the analysis in the **Number of Increments** edit box.
- Select the **Off** option from the **Intermediate Output** drop-down if you do not want to save output at each load increment. Select the **On** option to save output at each load increment. Select the **All** option from the **Intermediate Output** drop-down if you want to save all the outputs of analysis when nonlinear analysis bisects the load increment.
- Select the **On** option from the **Arc Length** drop-down if you are concerned about softening, buckling failure or material failure during the analysis. Note that activating this option takes more computing power for analysis.
- Click on the **Advanced Settings** button to set advanced nonlinear static parameters. The **Nonlinear Static Parameters** dialog box will be displayed; refer to Figure-25.

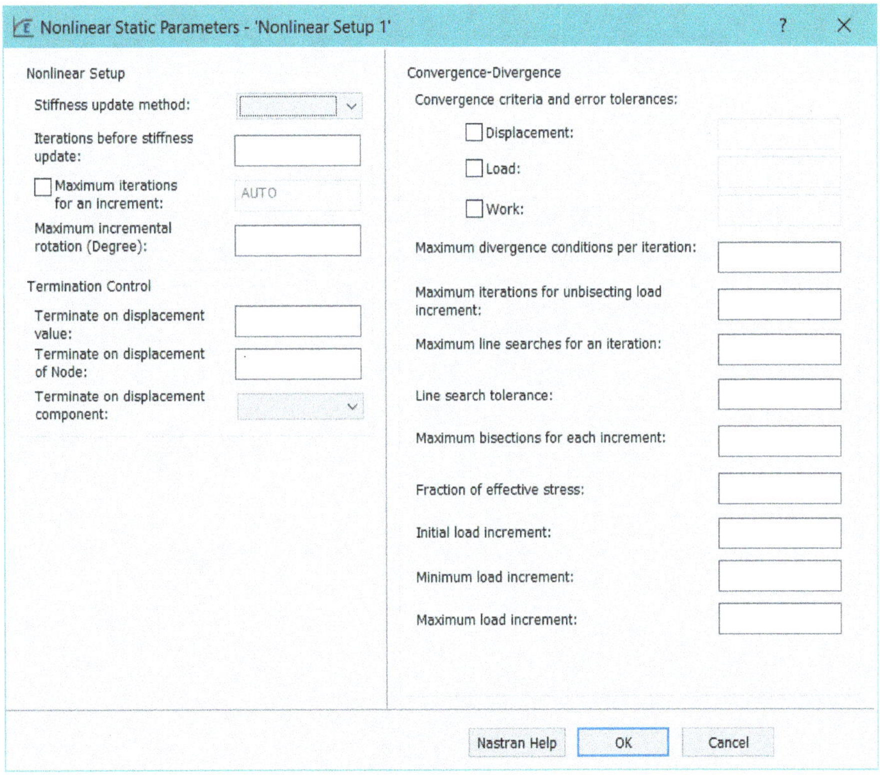

Figure-25. Nonlinear Static Parameters dialog box

- Set the desired parameters and click on the **OK** button. The **Nonlinear Setup** dialog box will be displayed again. Click on the **OK** button to apply the parameters and exit the dialog box.

- Apply material, constraint, and load as discussed earlier.
- Click on the **Run** button to run the analysis. Once the analysis is complete, the stages of result will be displayed as per the **Number of Increments** set in **Nonlinear Setup** dialog box; refer to Figure-26.

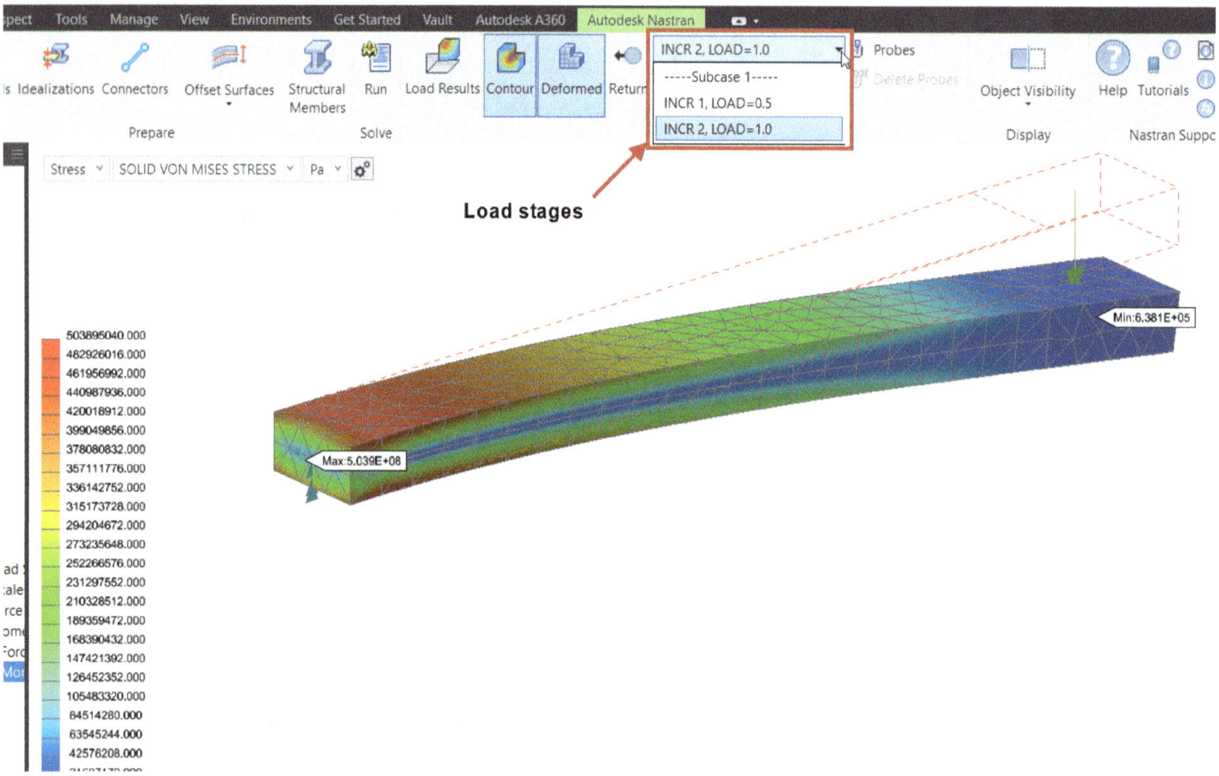

Figure-26. Result of nonlinear static analysis

The other options in **Autodesk Inventor Nastran** tab are same as discussed earlier. After performing the analysis, save the results and click on the **Finish Autodesk Inventor Nastran** tool from the **Exit** panel to exit.

PRACTICAL 1

Apply the conditions as shown in Figure-27 and perform linear static analysis on the model.

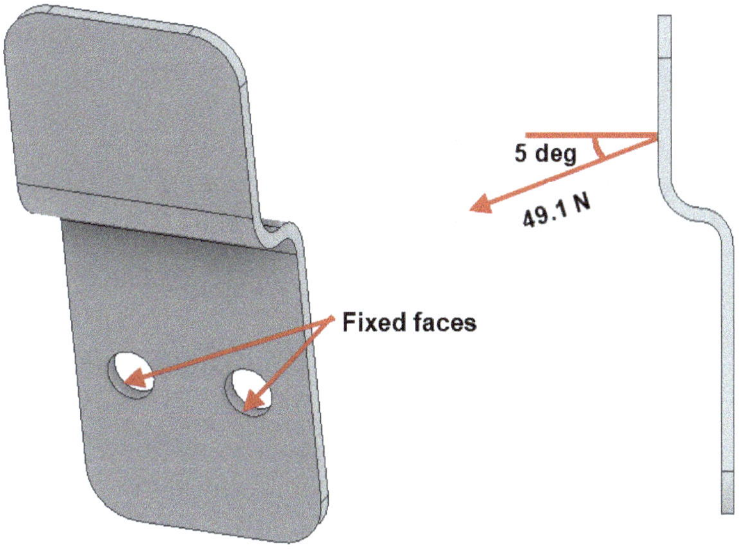

Figure-27. Practical 1

Opening Part and Starting Linear Static Analysis

- Open Practical 1 part file from Chapter 3 folder of Resources of this book and click on **Autodesk Inventor Nastran** tool from **Begin** panel of **Environments** tab in the **Ribbon**. The tools to perform linear static analysis will be displayed; refer to Figure-28.

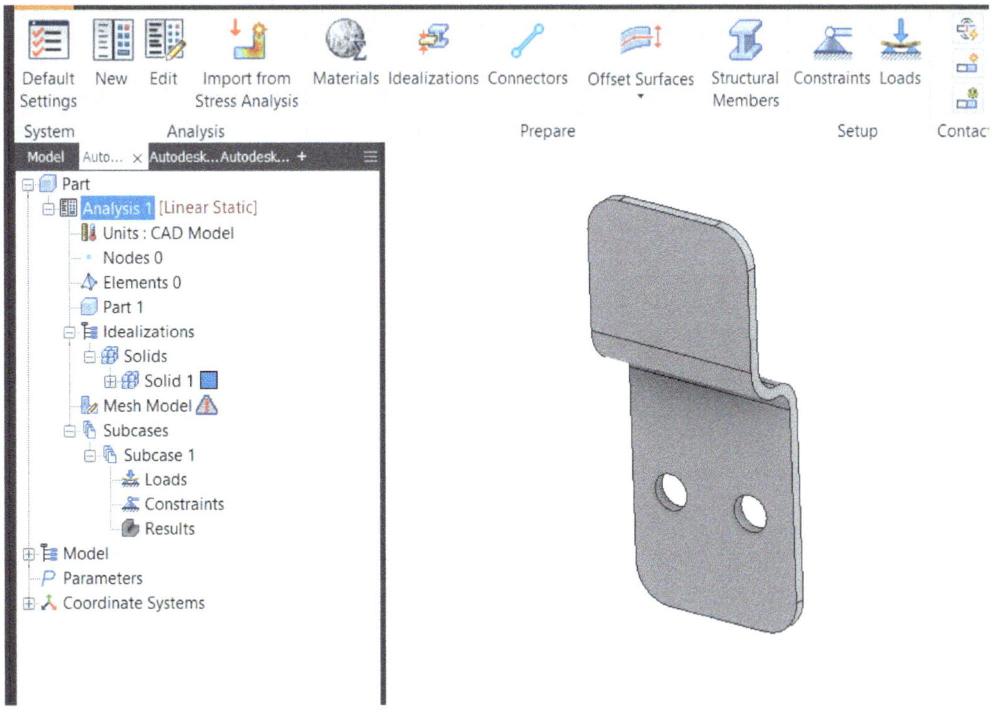

Figure-28. Linear static analysis options

Applying Material and Setting Idealization

- Click on **Materials** tool from the **Prepare** panel of **Autodesk Inventor Nastran** tab in the **Ribbon**. The **Material** dialog box will be displayed.

- Click on the **Select Material** button from the dialog box. The **Material DB** dialog box will be displayed.
- Select **Alloy Steel** material from the **Autodesk Material Library** node in the dialog box and click on the **OK** button.
- Click on the **OK** button from the **Material** dialog box. The material will be applied with solid idealization.
- Click on **Idealizations** tool from the **Prepare** panel in the **Autodesk Inventor Nastran** tab of **Ribbon**. The **Idealizations** dialog box will be displayed; refer to Figure-42.
- Select the desired color from the color section of the dialog box.

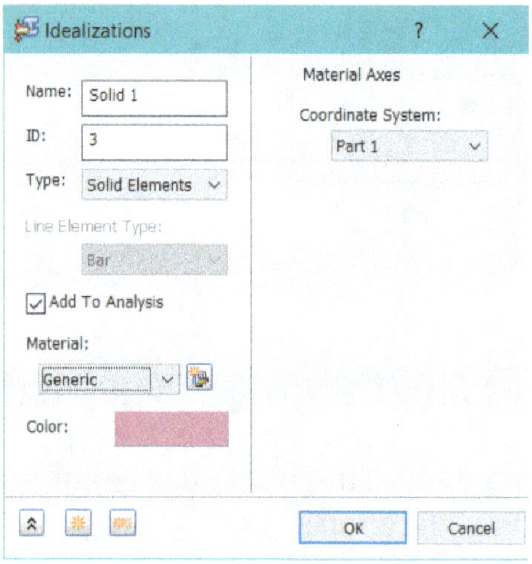

Figure-29. Idealizations dialog box

- Select the **Alloy Steel** option from the **Material** drop-down and click on the **OK** button.

Generating Mesh

- Click on **Mesh Settings** tool from the **Mesh** panel of **Autodesk Inventor Nastran** tab in the **Ribbon**. The **Mesh Settings** dialog box will be displayed.
- Specify element size as **3.5** in **Element Size** edit box.
- Click on **Generate Mesh** tool from the dialog box. The mesh will be generated; refer to Figure-30.

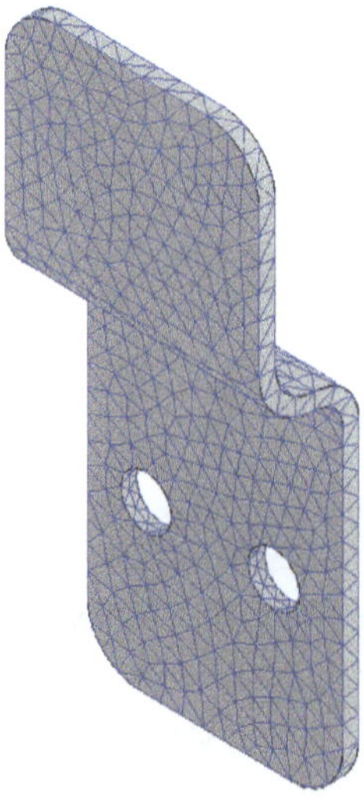

Figure-30. Mesh generated for Practical 1

- Click on the **OK** button from the **Mesh Settings** dialog box.

Applying Constraints

- Click on the **Constraints** tool from the **Setup** panel of **Autodesk Inventor Nastran** tab in the **Ribbon**. The **Constraint** dialog box will be displayed.
- Select the circular faces of holes to be constrained and click on the **Fixed** button; refer to Figure-31.

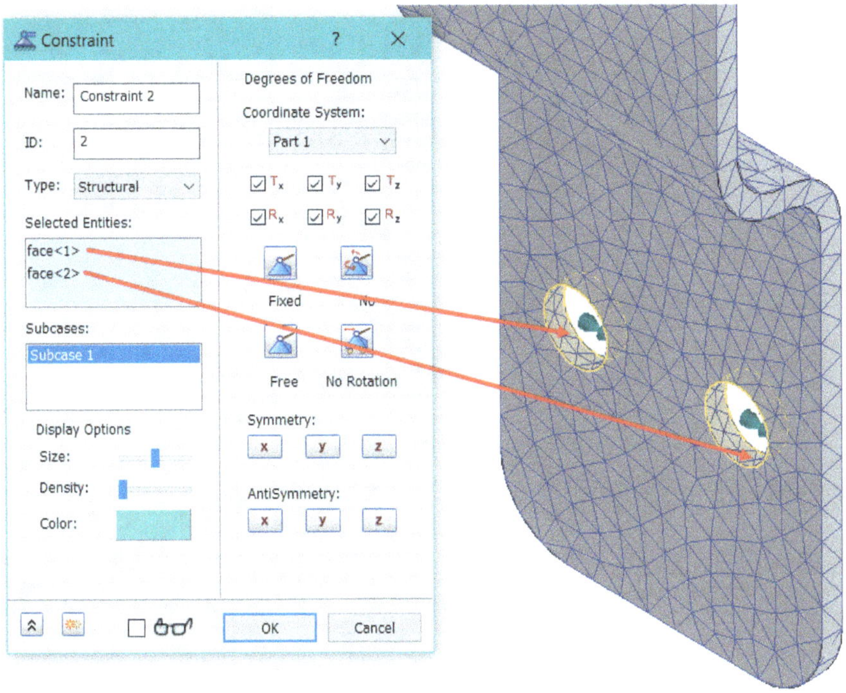

Figure-31. Faces selected for fixing circular faces

- Click on the **OK** button to apply constraint.

Applying Force

- Click on **Start 2D Sketch** tool from the **Sketch** panel of **Sketch** tab in the **Ribbon**. You will be asked to select a face/plane as sketching plane.
- Select the side face of the model; refer to Figure-32. The sketching environment will be displayed.

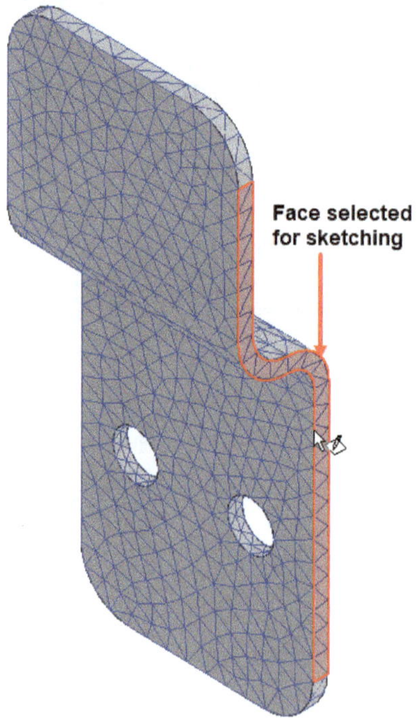

Figure-32. Face selected for sketching

- Click on the **Line** tool from the **Create** panel of **Sketch** tab in **Ribbon**. You will be asked to specify start point of the line.
- Create a line of 5 degree and 50 mm length as shown in Figure-33 and click on the **Finish Sketch** button from **Exit** panel.

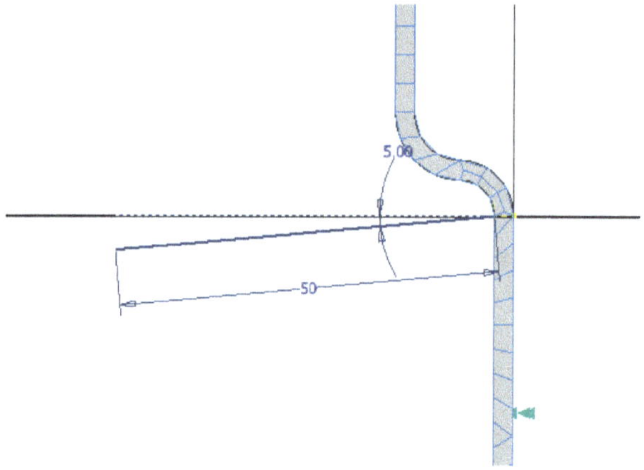

Figure-33. Line created for sketch

- Click on the **Loads** tool from the **Setup** panel of **Autodesk Inventor Nastran** tab in the **Ribbon**. The **Load** dialog box will be displayed.
- Select the **Force** option from the **Type** drop-down and select the upper face of model; refer to Figure-34.

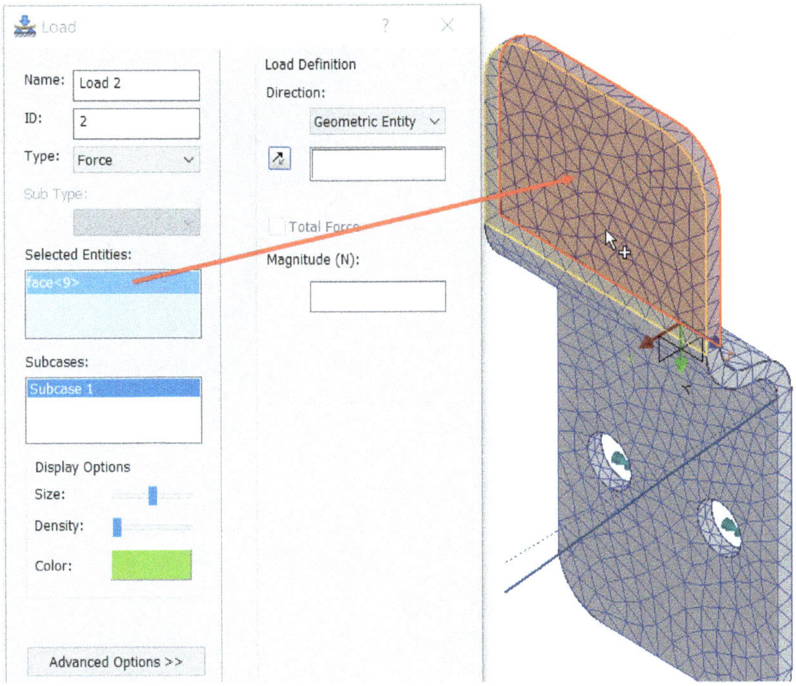

Figure-34. Face selected for force

- Select the **Geometric Entity** option from the **Direction** drop-down. The selection box below it will become active. Select the sketched line earlier created to define direction of force.
- Specify the value of force as **49.1** in the **Magnitude** edit box; refer to Figure-35.
- Click on the **OK** button from the dialog box.

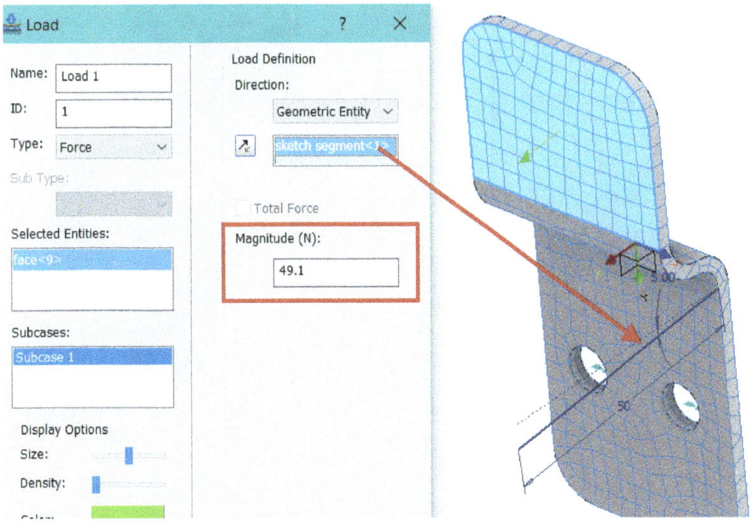

Figure-35. Applying force along selected line

Running Analysis

- Click on the **Run** button from the **Solve** panel of **Autodesk Inventor Nastran** tab in the **Ribbon**. Once the analysis is complete, the **Autodesk Inventor Nastran** information box will be displayed.
- Click on the **OK** button from the information box. The result of analysis will be displayed; refer to Figure-36.

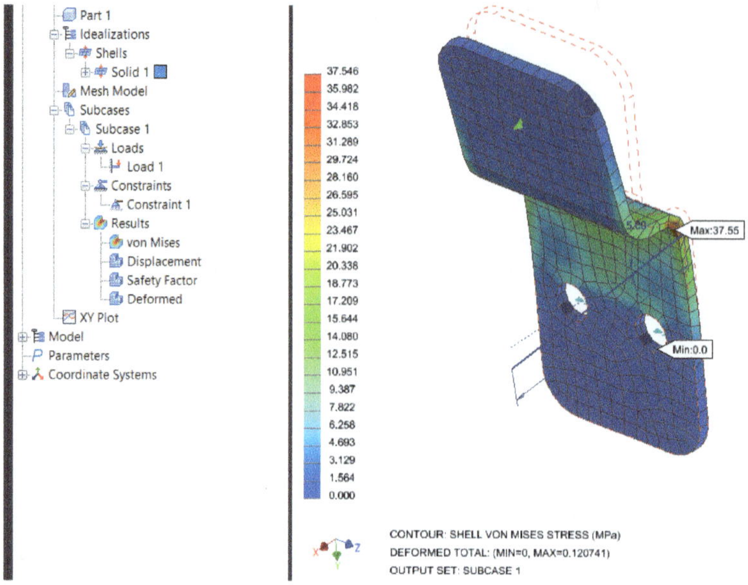

Figure-36. Result of practical 1

- Double-click on **Safety Factor** option under **Results** node to check the factor of safety; refer to Figure-37. Since, factor of safety is over 3 in critical areas of the model so our model is safe for operation. To be sure that the component will function properly, its factor of safety should be at least 3 at each critical location on the model. We can increase the strength of this part is by increasing thickness of part or by adding rib to part.

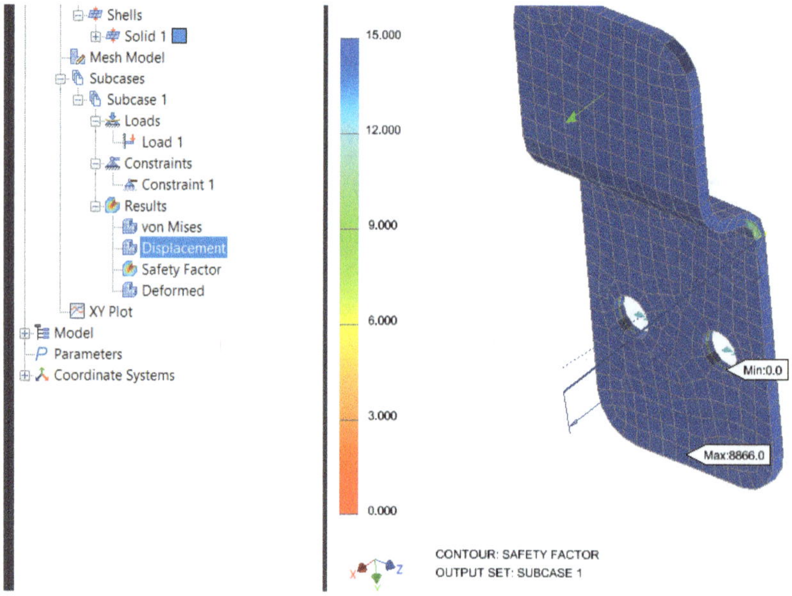

Figure-37. Factor of safety plot for practical 1

PRACTICAL 2

Perform linear static analysis on the model shown in Figure-38 with specified conditions.

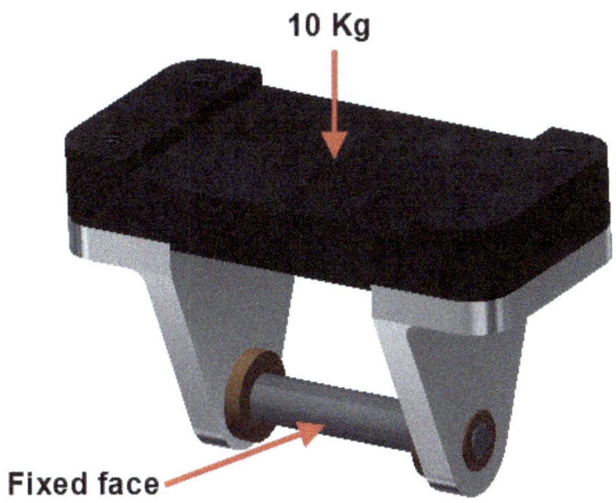

Figure-38. Practical 2

Opening Part and Starting Linear Static Analysis

Open Practical 2 assembly file from Chapter 3 folder of Resources of this book and click on **Autodesk Inventor Nastran** tool from **Begin** panel of **Environments** tab in the **Ribbon**. The tools to perform linear static analysis will be displayed.

Note that solid idealization and materials are already applied to the model.

Applying Constraints

- Click on the **Constraints** tool from the **Setup** panel of **Autodesk Inventor Nastran** tab in the **Ribbon**. The **Constraint** dialog box will be displayed.
- Select the round face of the bolt and click on the **Fixed** button; refer to Figure-39.

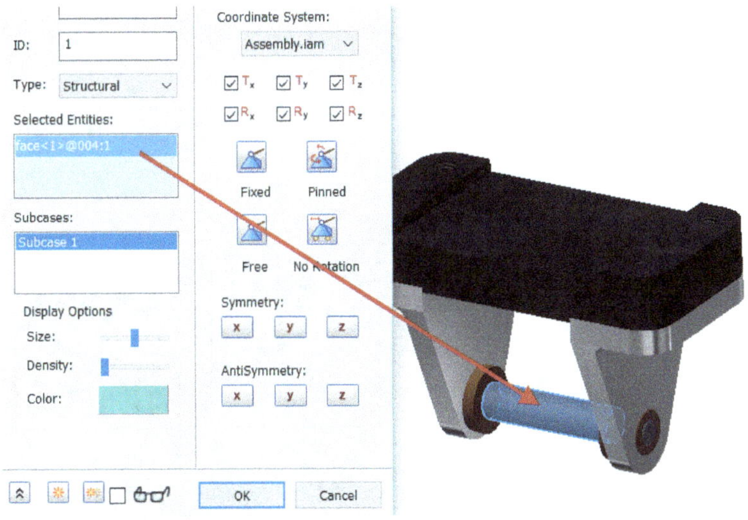

Figure-39. Applying constraint

- After setting the desired parameters, click on the **OK** button.

Applying Load

- Click on the **Loads** tool from the **Setup** panel of **Autodesk Inventor Nastran** tab in the **Ribbon**. The **Loads** dialog box will be displayed.
- Select the top face of the model and specify the force value as **98.1** in the edit box; refer to Figure-40.

Figure-40. Specifying weight on model

- Click on the **OK** button from the dialog box.

Running the Analysis

- Click on the **Mesh Settings** tool from the **Mesh** panel of the **Ribbon**. The **Mesh Settings** dialog box will be displayed.
- Set the element size as **5** mm in the **Element Size** edit box and click on the **Generate Mesh** button. The mesh will be generated; refer to Figure-41.

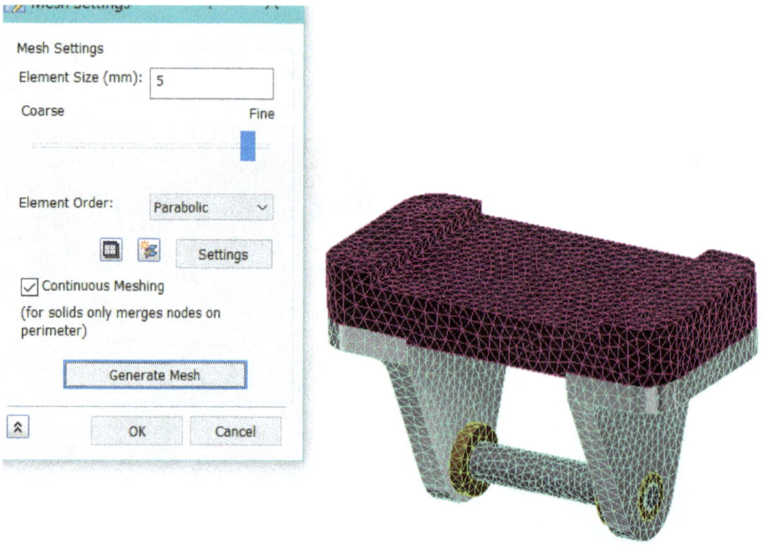

Figure-41. Mesh generated for practical 2

- Click on the **OK** button from the dialog box.

- Click on the **Run** button from the **Solve** panel of **Ribbon.** Once the analysis is complete, the **Autodesk Inventor Nastran** information box will be displayed.
- Click on the **OK** button from the information box. Results of analysis will be displayed; refer to Figure-42.

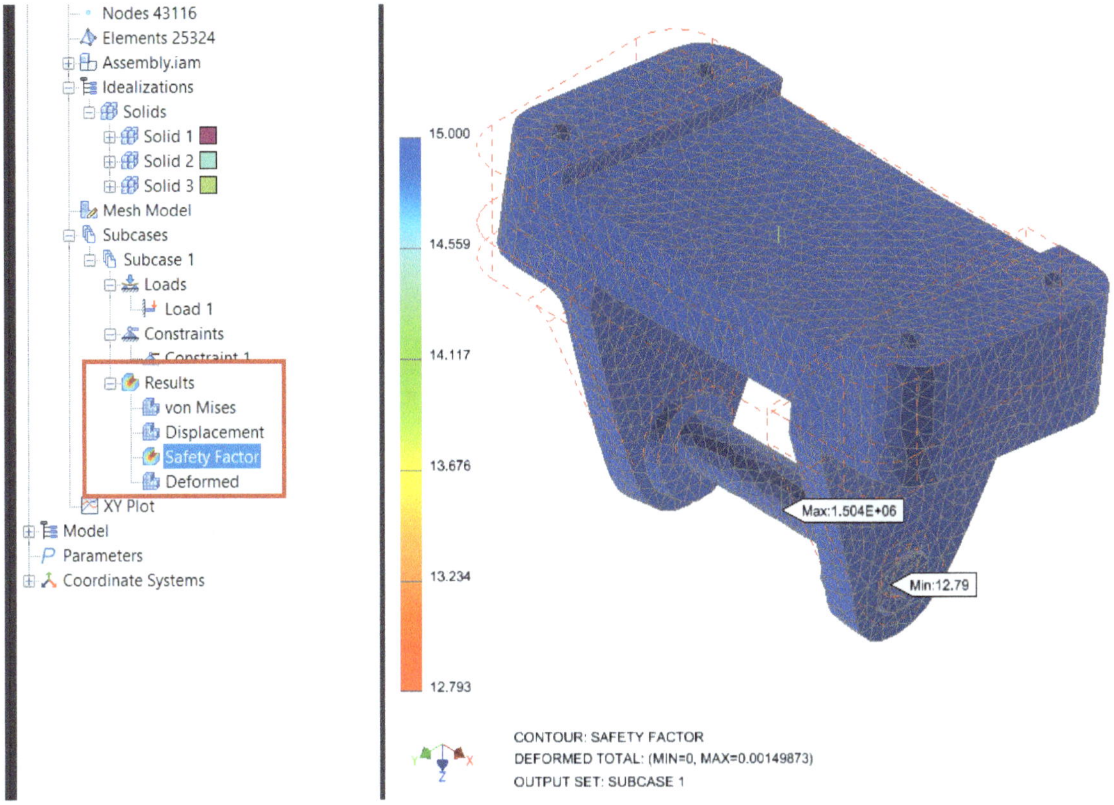

Figure-42. Results of analysis for Practical 2

Check the results and modify the model accordingly.

For Student Notes

Chapter 4

Normal Modes Analysis

Topics Covered

The major topics covered in this chapter are:

- ***Introduction.***
- ***Starting Normal Modes Analysis***
- ***Specifying Modal Setup Parameters***
- ***Applying Materials, Loads, and Constraints***
- ***Applying Idealizations***
- ***Setting Mesh and other parameters***
- ***Running Analysis and Plotting Results***
- ***Performing Prestress Normal Modes Analysis***

INTRODUCTION

Every structure has the tendency to vibrate at certain frequencies, called **natural or resonant frequencies**. Each natural frequency is associated with a certain shape, called **mode shape**, that the model tends to assume when vibrating at that frequency. To check these frequencies, we perform Normal Modes analysis.

When a structure is properly excited by a dynamic load with a frequency that coincides with one of its natural frequencies, the structure undergoes large displacements and stresses. This phenomenon is known as **resonance**. For undamped systems, resonance theoretically causes infinite motion. **Damping**, however, puts a limit on the response of the structures due to resonant loads.

A real model has an infinite number of natural frequencies. However, a finite element model has a finite number of natural frequencies that are equal to the number of degrees of freedom considered in the model. Only the first few modes are needed for most purposes.

If your design is subjected to dynamic environments, static studies cannot be used to evaluate the response. Frequency studies can help you design vibration isolated systems by avoiding resonance in specific frequency band. They also form the basis for evaluating the response of linear dynamic systems where the response of a system to a dynamic environment is assumed to be equal to the summation of the contributions of the modes considered in the analysis.

Note that resonance is desirable in the design of some devices. For example, resonance is required in guitars and violins.

The natural frequencies and corresponding mode shapes depend on the geometry, material properties, and support conditions. The computation of natural frequencies and mode shapes are known as modal, frequency, and normal mode analyses.

When building the geometry of a model, you usually create it based on the original (undeformed) shape of the model. Some loads, like the structure's own weight, are always present and can cause considerable effects on the shape of the structure and its modal properties. In many cases, this effect can be ignored because the induced deflections are small.

Loads affect the modal characteristics of a body. In general, compressive loads decrease resonant frequencies and tensile loads increase them. This fact is easily demonstrated by changing the tension on a violin string. The higher the tension, the higher the frequency (tone).

You do not need to define any loads for a frequency study but if you do, their effect will be considered. By having evaluated natural frequencies of a structure's vibrations at the design stage, you can optimize the structure with the goal of meeting the frequency vibro-stability condition. To increase natural frequencies, you would need to add rigidity to the structure and (or) reduce its weight. For example, in the case of a slender object, the rigidity can be increased by reducing the length and increasing the thickness of the object. To reduce a part's natural frequency, you should, on the contrary, increase the weight or reduce the object's rigidity.

In Autodesk Inventor Nastran, we can perform two types of Normal Mode analyses viz. Normal Modes analysis and Prestress Normal Modes Analysis.

PERFORMING NORMAL MODES ANALYSIS

The procedure to perform Normal Modes analysis is given next.

- Open the part on which you want to perform normal mode analysis and start Autodesk Inventor Nastran environment.
- Click on the **New** button from the **Analysis** panel of **Autodesk Inventor Nastran** tab in **Ribbon**. The **Analysis** dialog box will be displayed.
- Select the **Normal Modes** option from the **Type** drop-down in **Analysis** dialog box.
- Set the other options as discussed earlier and click on the **OK** button. The options in **Model Tree** will be displayed as shown in Figure-1.

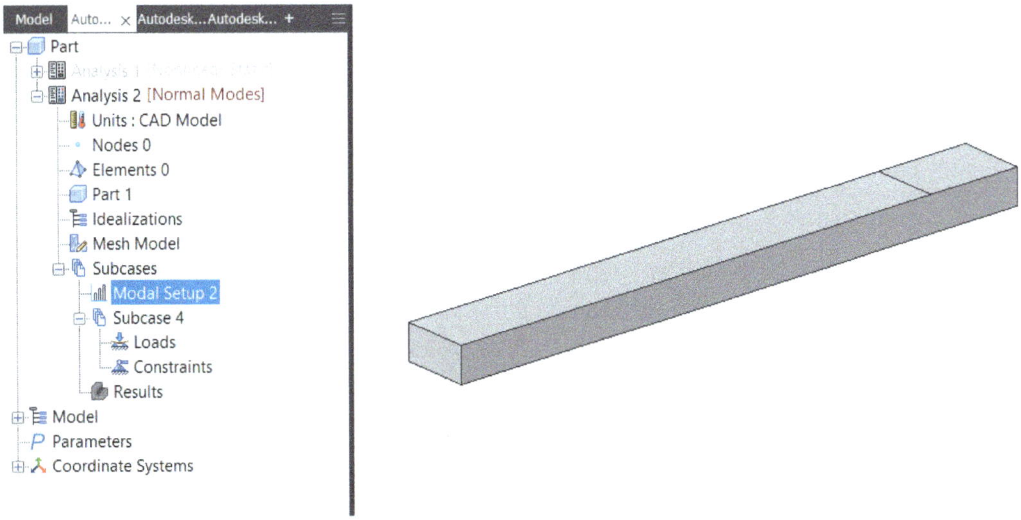

Figure-1. Options for Normal Modes analysis

Modal Setup Options

- Double-click on **Modal Setup** option from **Subcases** node to define parameters related to modal analysis. The **Modal Setup** dialog box will be displayed; refer to Figure-2.

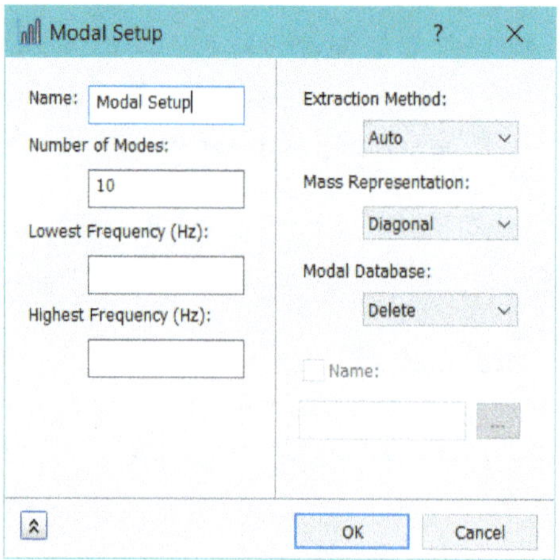

Figure-2. Modal Setup dialog box

- Specify desired name for model setup parameter in the **Name** edit box.
- Specify the number of natural shapes to be found during the analysis in **Number of Modes** edit box.
- Specify the desired lower and higher limit of frequencies within which natural modes will be calculated in **Lowest Frequency** and **Highest Frequency** edit boxes, respectively. Note that generally, the values specified here are those frequencies within which your machine will be working. Using this analysis, you want to make sure that the part/assembly to be designed does not have any natural frequency in specified range.
- Select the desired option from the **Extraction Method** drop-down to define how natural frequencies will be calculated. Select the **Lanczos Iteration** option for larger problems as it is replaces the older values of each equation with new values as they are solved in same iteration which in turn reduces the number of total iterations required for convergence. Select the Subspace Iteration option from the drop-down when you need to perform analysis with parallel iterations. Although, the number of iterations required for **Subspace** method is higher but when there is a chance of getting error with Lanczos method then you should use this method.
- Select the desired option from the **Mass Representation** drop-down to define how mass matrices will be formed for analysis.
- Select the desired option from the **Modal Database** drop-down to define storage and retrieval of modal data like eigenvalues and eigenvectors. Note that selecting the **Element Type Default** option will use coupled mass matrix when rigid elements are specified in the model.
- If you have selected any option other than **Delete** from the **Modal Database** drop-down then options to define location of database will be displayed. Set the desired location for database storage. The eigenvalues and eigenvectors are stored in the database.
- After setting desired values, click on the **OK** button.

- Although it is not mandatory to apply load in Normal Mode analysis but if you add high value load then it will be counted in analysis.

- Apply the desired constraint, set idealization, and click on the **Run** button to perform analysis. The results will be displayed with list of mode frequencies; refer to Figure-3.

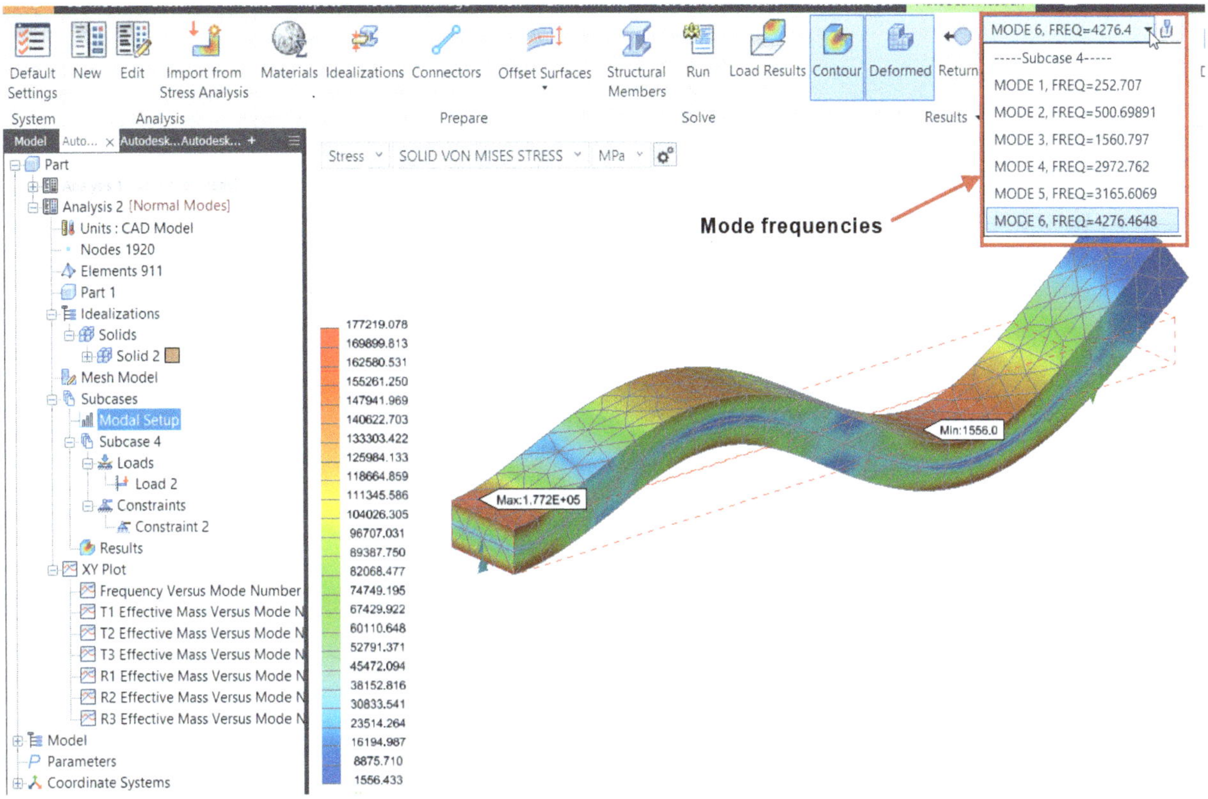

Figure-3. Normal Modes analysis result with mode frequencies

- Double-click on desired result from the **XY Plot** node to check the plots.
- Save the analysis results and exit the **Autodesk Inventor Nastran** environment.

PERFORMING PRESTRESS NORMAL MODES ANALYSIS

Prestress Normal Modes analysis is performed on models which are already under stress due to continuous load. The procedure to perform Prestress Normal Modes analysis is given next.

- Open the desired model and start **Autodesk Inventor Nastran** environment.
- Click on the **New** button from the **Analysis** panel of **Autodesk Inventor Nastran** tab in the **Ribbon**. The **Analysis** dialog box will be displayed.
- Select the **Prestress Normal Modes** option from the **Type** drop-down and set the other options as required.
- Click on the **OK** button from the dialog box. The load sub cases for the analysis will be displayed; refer to Figure-4.

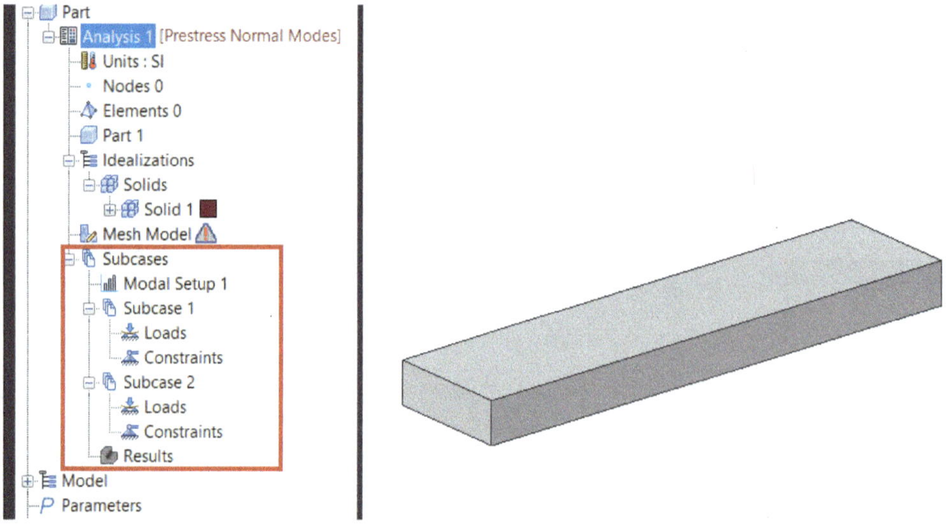

Figure-4. Subcase options for prestress normal modes analysis

- Set the desired parameters for normal modes in the **Modal Setup** dialog box as discussed earlier.
- Set the desired load and constraint in Subcase 1 which will be set as prestress load.
- Set the desired load and constraint in Subcase 2 which will be applied as additional load.
- Apply the desired material and set the idealization.
- Click on the **Run** button. The analysis results will be displayed; refer to Figure-5.

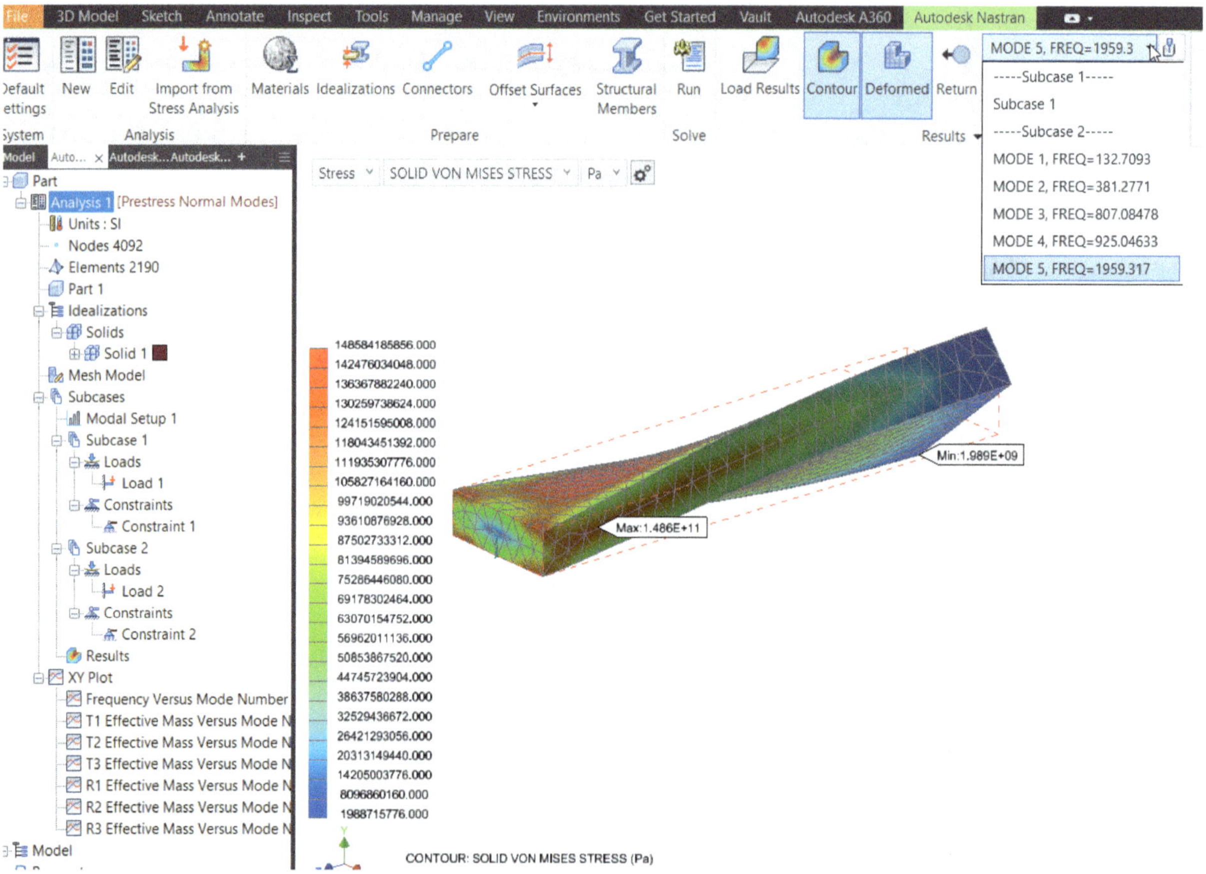

Figure-5. Prestress normal mode analysis result

- Double-click on desired result under **XY Plot** node to check the results.

PRACTICAL

Find out natural frequency of tuning fork as shown in Figure-6.

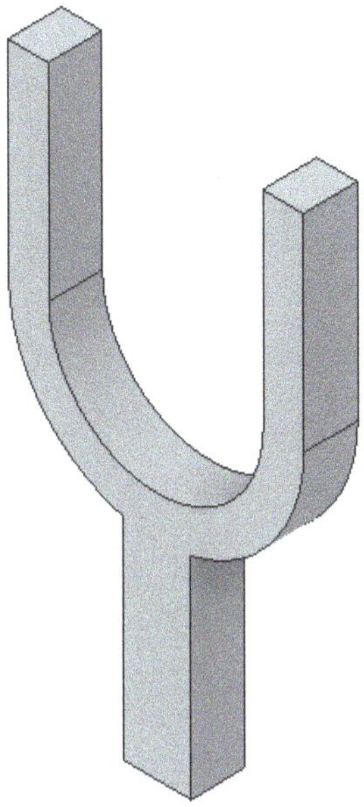

Figure-6. Tuning fork

Opening Part and Starting Autodesk Inventor Nastran

- Open the Tuning Fork part file from Chapter 4 folder of Resource folder for the book.
- Click on **Autodesk Inventor Nastran** tool from the **Begin** panel of **Autodesk Inventor Nastran** tab in the **Ribbon**. Linear static analysis will be activated automatically.
- Click on the **Edit** button from the **Analysis** panel of **Autodesk Inventor Nastran** tab in the **Ribbon**. The **Analysis** dialog box will be displayed.
- Select the **Normal Modes** option from the **Type** drop-down. The options for natural frequency analysis will be displayed.
- Click on the **OK** button from the dialog box.

Applying Material and Setting Idealization

- Click on the **Materials** tool from the **Prepare** panel of **Autodesk Inventor Nastran** tab. The **Material** dialog box will be displayed.
- Click on the **Select Material** button from the **Material** dialog box. The **Material DB** dialog box will be displayed.
- Select the **Stainless Steel** option from the **Inventor Material Library** node of **Material DB** dialog box and click on the **OK** button. The properties of selected material will be displayed in the **Material** dialog box.
- Click on the **OK** button from the **Material** dialog box to apply material.
- Double-click on **Solid 1** option from **Solids** node of **Idealizations** node. The **Idealizations** dialog box will be displayed.
- Select the **Stainless Steel** option from the **Material** drop-down and click on the **OK** button.

Applying Constraints and Generating Mesh

- Click on the **Constraints** tool from the **Setup** panel of **Autodesk Inventor Nastran** tab in the **Ribbon**. The **Constraint** dialog box will be displayed.
- Select the bottom side faces of model as shown in Figure-7 and click on the **Fixed** button from the dialog box.

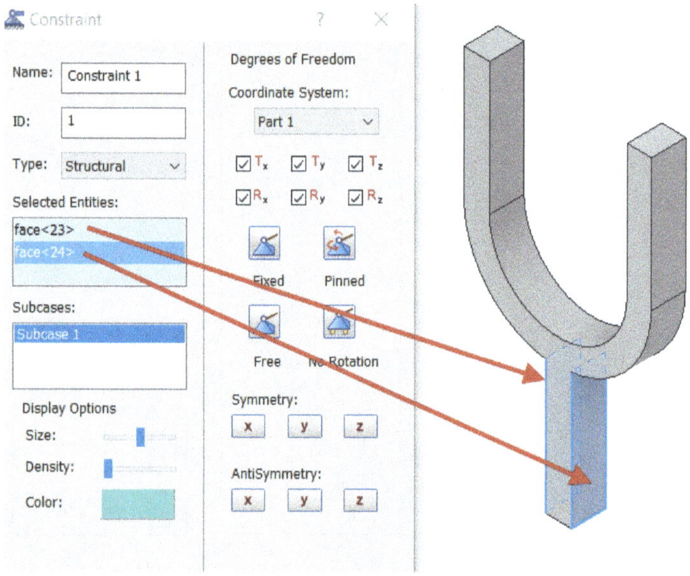

Figure-7. Faces selected for constraining

- Click on the **OK** button from the dialog box.
- Click on the **Mesh Settings** tool from the **Mesh** panel of **Autodesk Inventor Nastran** tab in the **Ribbon**. The **Mesh Settings** dialog box will be displayed.
- Specify the mesh element size as **15** in **Element Size (mm)** edit box.
- Click on the **Generate Mesh** button from the dialog box. The mesh will be generated; refer to Figure-8.

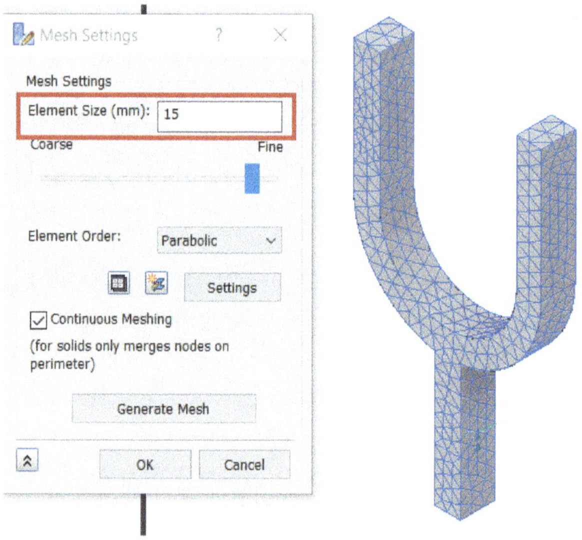

Figure-8. Mesh generated for tuning fork

- Click on the **OK** button from the dialog box.

Applying Modal Setup Parameters

- Double-click on **Modal Setup 1** option from the **Subcases** node in the **Model Tree**. The **Modal Setup** dialog box will be displayed.

- Specify the lowest frequency as **10** Hz and highest frequency as **1500** Hz in respective edit boxes of the dialog box.
- Set the other options as shown in Figure-9 and click on the **OK** button.

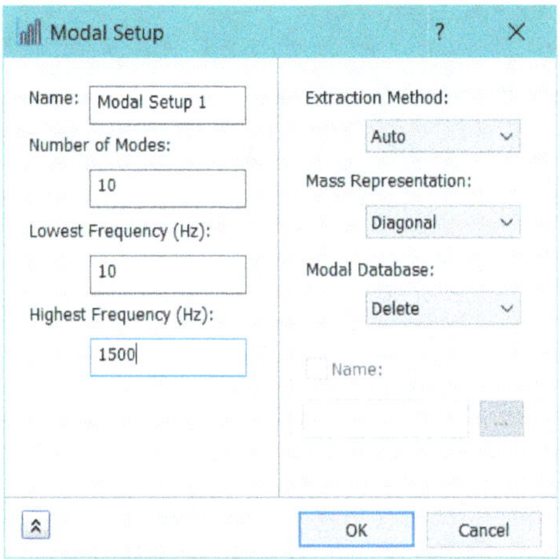

Figure-9. Modal Setup dialog box options to be specified

Running Analysis

- Click on the **Run** tool from the **Solve** panel of **Autodesk Inventor Nastran** tab in the **Ribbon**. Once the analysis is complete, **Autodesk Inventor Nastran** information box will be displayed.
- Click on the **OK** button from the information box. Results of analysis will be displayed; refer to Figure-10.

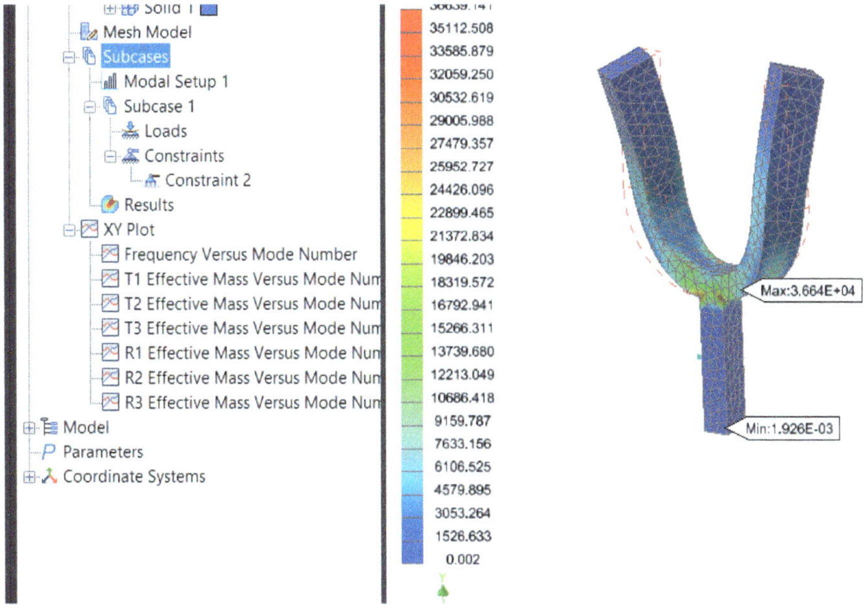

Figure-10. Result of normal analysis

Double-click on **Frequency Versus Mode Number** option from the **XY Plot** node. The **XY Plot** will be displayed with mode numbers and respective natural frequencies; refer to Figure-11.

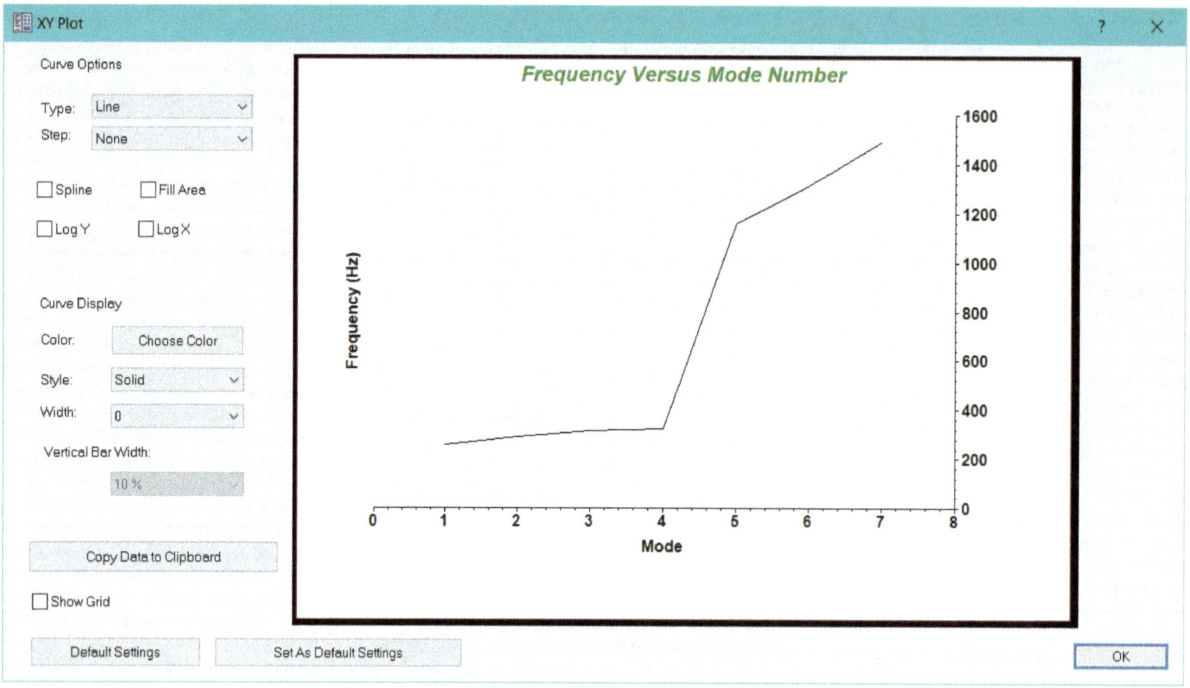

Figure-11. XY Plot for natural frequencies

Check the results and click on the **OK** button. Click on the **Finish Autodesk Inventor Nastran** tool from the **Exit** panel and save the model.

PRACTICE

Perform Normal mode analysis with prestress on the model as shown in Figure-12. The model used for this practice is same as discussed in Practical above.

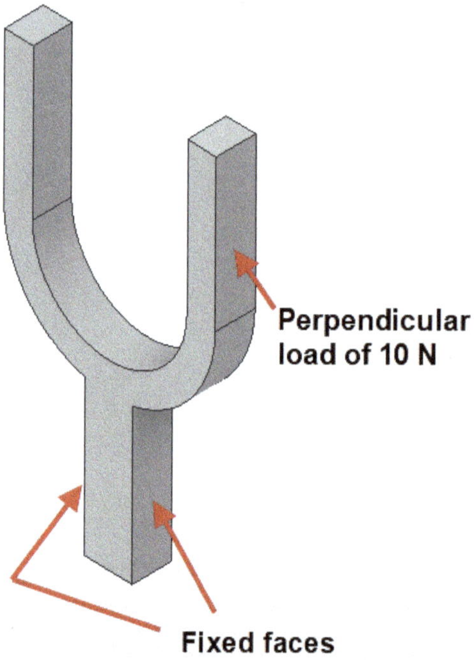

Figure-12. Tuning fork for prestress analysis with normal mode

Chapter 5

Buckling Analysis

Topics Covered

The major topics covered in this chapter are:

- ***Introduction.***
- ***Performing Linear Buckling Analysis***
- ***Performing Nonlinear Buckling Analysis***

INTRODUCTION

Slender models tends to buckle under axial loading. Buckling is defined as the sudden deformation which occurs when the stored membrane (axial) energy is converted into bending energy with no change in the externally applied loads. Mathematically, when buckling occurs, the stiffness becomes singular. The Linearized buckling approach, used here, solves an eigenvalue problem to estimate the critical buckling factors and the associated buckling mode shapes.

A model can buckle in different shapes under different levels of loading. The shape the model takes while buckling is called the buckling mode shape and the loading is called the critical or buckling load. Buckling analysis calculates a number of modes as requested in the Buckling dialog. Designers are usually interested in the lowest mode (mode 1) because it is associated with the lowest critical load. When buckling is the critical design factor, calculating multiple buckling modes helps in locating the weak areas of the model. The mode shapes can help you modify the model or the support system to prevent buckling in a certain mode.

A more vigorous approach to study the behavior of models at and beyond buckling requires the use of nonlinear design analysis codes.

In a laymen's language, if you press down on an empty soft drink can with your hand, not much will seem to happen. If you put the can on the floor and gradually increase the force by stepping down on it with your foot, at some point it will suddenly squash. This sudden scrunching is known as "buckling."

Models with thin parts tend to buckle under axial loading. Buckling can be defined as the sudden deformation, which occurs when the stored membrane (axial) energy is converted into bending energy with no change in the externally applied loads. Mathematically, when buckling occurs, the total stiffness matrix becomes singular.

In the normal use of most products, buckling can be catastrophic if it occurs. The failure is not one because of stress but geometric stability. Once the geometry of the part starts to deform, it can no longer support even a fraction of the force initially applied. The worst part about buckling for engineers is that buckling usually occurs at relatively low stress values for what the material can withstand. So they have to make a separate check to see if a product or part thereof is okay with respect to buckling.

Slender structures and structures with slender parts loaded in the axial direction buckle under relatively small axial loads. Such structures may fail in buckling while their stresses are far below critical levels. For such structures, the buckling load becomes a critical design factor. Stocky structures, on the other hand, require large loads to buckle, therefore buckling analysis is usually not required.

Buckling almost always involves compression. In civil engineering, buckling is to be avoided when designing support columns, load bearing walls and sections of bridges which may flex under load. For example an I-beam may be perfectly "safe" when considering only the maximum stress, but fail disastrously if just one local spot of a flange should buckle! In mechanical engineering, designs involving thin parts in flexible structures like airplanes and automobiles are susceptible to buckling. Even though stress can be very low, buckling of local areas can cause the whole structure to collapse by a rapid series of 'propagating buckling'.

Buckling analysis calculates the smallest (critical) loading required for buckling a model. Buckling loads are associated with buckling modes. Designers are usually interested in the lowest mode because it is associated with the lowest critical load. When buckling is the critical design factor, calculating multiple buckling modes helps in locating the weak areas of the model. This may prevent the occurrence of lower buckling modes by simple modifications.

PERFORMING LINEAR BUCKLING ANALYSIS

Slender parts and assemblies with slender components that are loaded in the axial direction buckle under relatively small axial loads. Such structures can fail due to buckling while the stresses are far below critical levels. For such structures, the buckling load becomes a critical design factor. Buckling analysis is usually not required for bulky structures as failure occurs earlier due to high stresses. The procedure to use buckling analysis in Autodesk Inventor Nastran is given next.

- Open the part on which you want to perform buckling analysis and start Autodesk Inventor Nastran environment.
- Click on the **Edit** tool from the **Analysis** panel of **Autodesk Inventor Nastran** tab in **Ribbon**. The **Analysis** dialog box will be displayed.
- Select the **Linear Buckling** option from the **Type** drop-down and set the desired parameters in **Output Controls** tab.
- Click on the **Options** tab in the dialog box. The options will be displayed as shown in Figure-1.

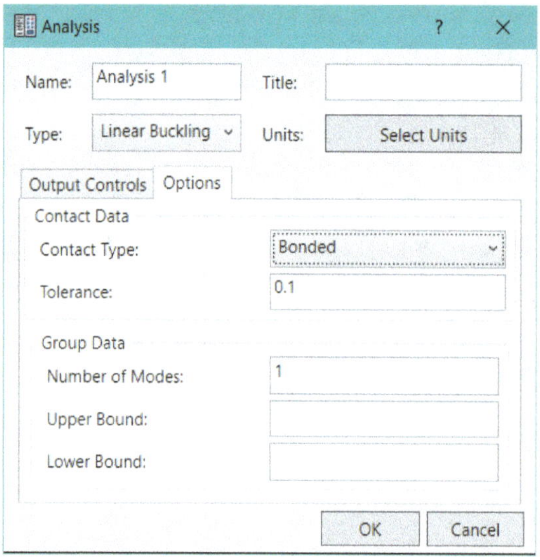

Figure-1. Buckling analysis options

- Set the desired number of buckling modes you want to check in the **Number of Modes** edit box.
- Specify the desired stress values for upper and lower limits in the **Upper Bound** and **Lower Bound** edit boxes.
- Click on the **OK** button from the dialog box. The options related to linear buckling analysis will be displayed.
- Apply the desired value of load and set the desired constraint.
- Click on the **Run** button from the **Solve** panel. The buckling analysis result will be displayed; refer to Figure-2.

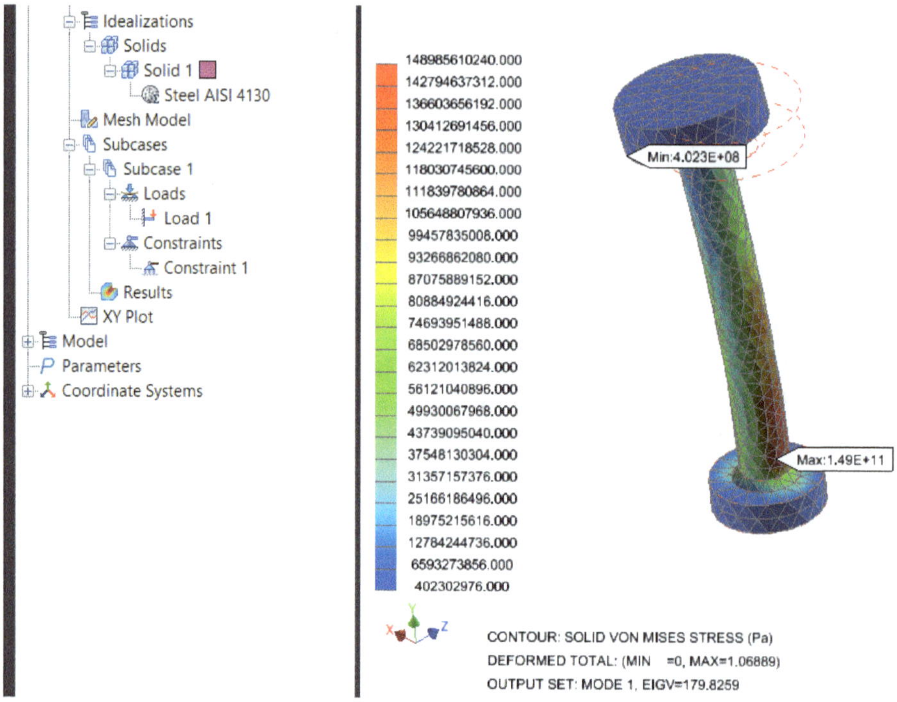

Figure-2. Buckling analysis result

PERFORMING NON-LINEAR BUCKLING ANALYSIS

The non-linear buckling analysis is used when there is large displacement in model due to buckling and you are concerned about the shape attained by object due to buckling. In other words, if you want the object to buckle under load and want to check its shaped attained due to buckling then you should use the non-linear buckling analysis. The procedure to use buckling analysis in Autodesk Inventor Nastran is given next.

- Open the part and start Autodesk Inventor Nastran.
- Click on the **New** button from the **Analysis** panel of **Autodesk Inventor Nastran** tab in **Ribbon**. The **Analysis** dialog box will be displayed.
- Select the **Nonlinear Buckling** option from the **Type** drop-down and set the desired parameters in **Output Controls** tab of the dialog box.
- Set the desired number of buckling modes, upper bound and lower bound for the analysis. Make sure so select the **On** option for **Large Displacements** drop-down if there is going to be large deformation.
- Click on the **OK** button from the dialog box after setting the desired parameters. The options for nonlinear analysis will be displayed; refer to Figure-3.

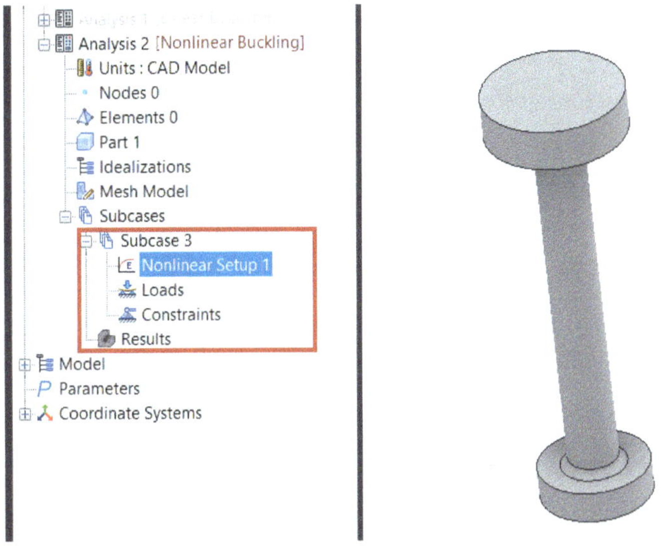

Figure-3. Nonlinear buckling analysis subcase

- Double-click on the **Nonlinear Setup 1** option from the **Subcase** node. The **Nonlinear Setup** dialog box will be displayed. Set the options as discussed earlier and click on the **OK** button.
- The other options are same as discussed for linear buckling analysis. Set the desired idealization, load, and constraint.
- After setting the desired parameters, click on the **Run** button. Once the analysis is complete, the result of analysis will be displayed; refer to Figure-4.

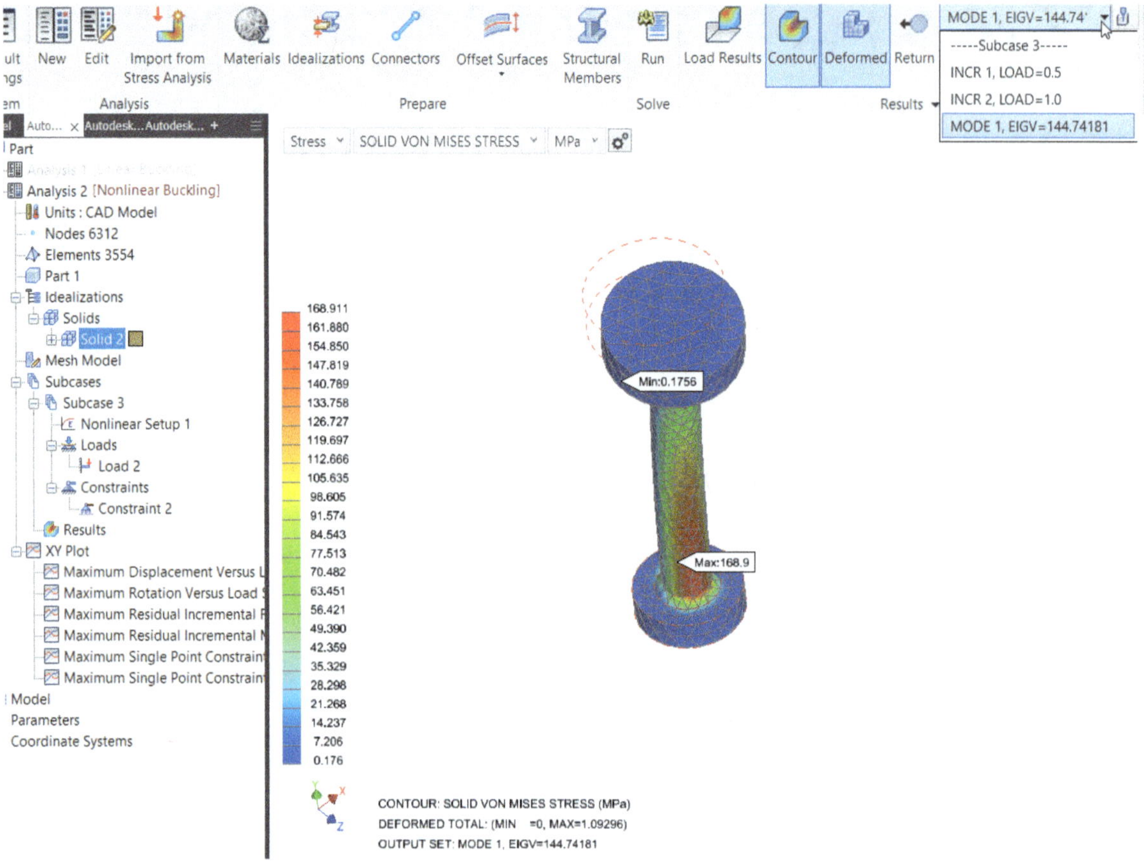

Figure-4. Result of nonlinear buckling analysis

- Double-click on the desired result under XY Plot node.

PRACTICAL 1

Perform buckling analysis on the model as shown in Figure-5.

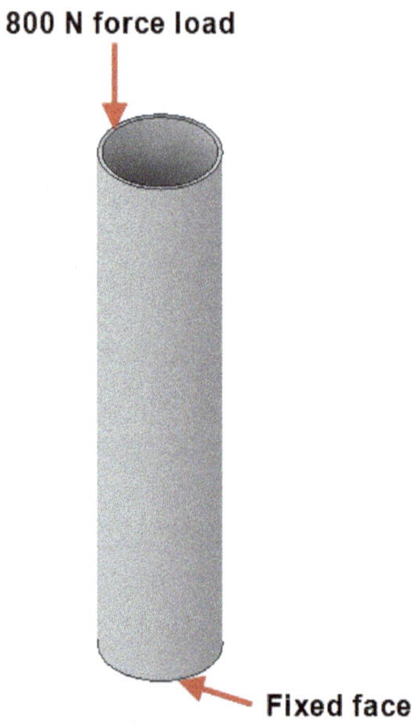

Figure-5. Parameters for buckling analysis

Opening Part and Starting Buckling Analysis

- Open the Practical for buckling analysis part from Chapter 5 folder of Resources for Autodesk Inventor Nastran.
- Click on the **Autodesk Inventor Nastran** tool from the **Begin** panel of **Environments** tab in the **Ribbon**. The linear static analysis will start automatically.
- Click on the **Edit** button from the **Analysis** panel of **Autodesk Inventor Nastran** tab in the **Ribbon**. The **Analysis** dialog box will be displayed.
- Select the **Linear Buckling** option from the **Type** drop-down and click on the **Options** tab. Specify the number of modes as **5** in the **Number of Modes** edit box; refer to Figure-6.

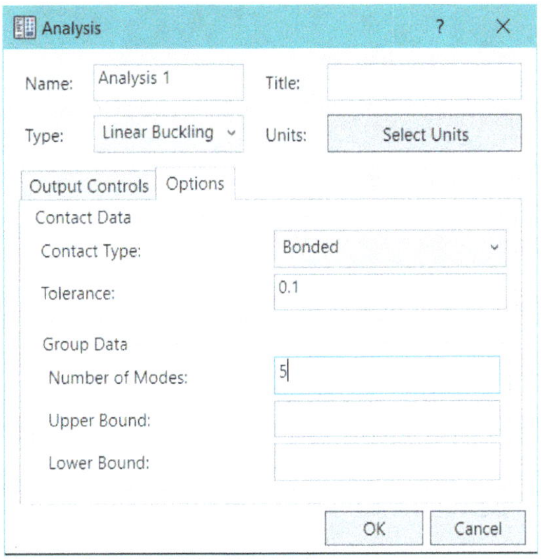

Figure-6. Specifying modes for buckling analysis

- Click on the **OK** button from the dialog box. The options to perform buckling analysis will be displayed.

Applying Material and Setting Idealization

- Click on the **Materials** tool from the **Prepare** panel of **Autodesk Inventor Nastran** tab in the **Ribbon**. The **Material** dialog box will be displayed.
- Select the **Stainless Steel** material from the dialog box and click on the **OK** button from the **Material** dialog box.
- Double-click on **Solid 1** option from the **Solids** node of **Idealizations** node in **Model Tree**. The **Idealizations** dialog box will be displayed.
- Select **Stainless Steel** option from the **Material** drop-down and click on the **OK** button.

Applying Constraint and Load

- Click on the **Constraint** tool from the **Setup** panel of **Autodesk Inventor Nastran** tab. The **Constraint** dialog box will be displayed.
- Select the bottom flat face of the model and click on the **Fixed** button; refer to Figure-7.

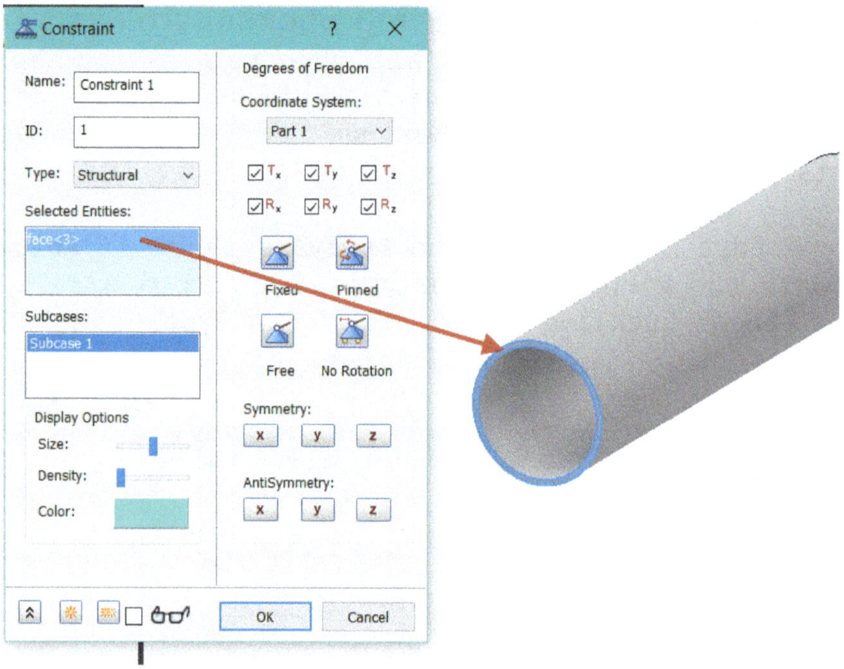

Figure-7. Face selected for fixing

- Click on the **OK** button from the dialog box.
- Click on the **Loads** tool from the **Setup** panel of **Autodesk Inventor Nastran** tab in the **Ribbon**. The **Load** dialog box will be displayed.
- Select the top face of model and select **Normal of Surface** option from the **Direction** drop-down.
- Specify the value of force as **-800** N in the **Magnitude** edit box; refer to Figure-8.

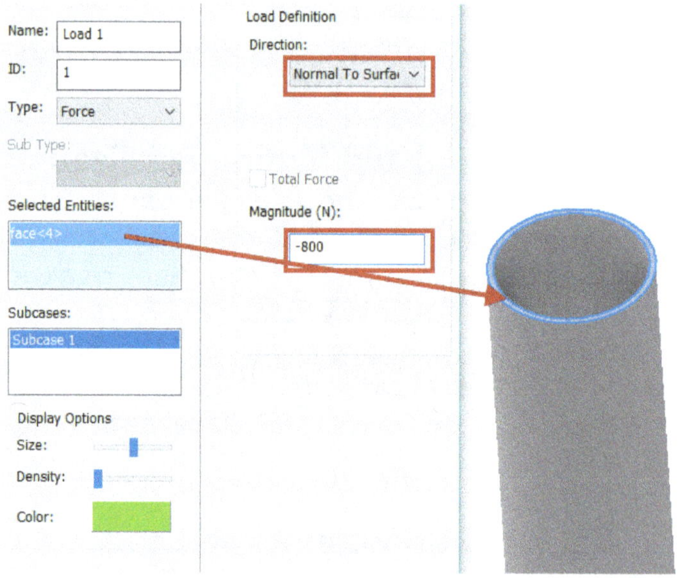

Figure-8. Applying load

- After setting desired parameters, click on the **OK** button from the dialog box.

Running Analysis

- Click on the **Mesh Settings** tool from the **Mesh** panel of **Autodesk Inventor Nastran** tab in the **Ribbon**. The **Mesh Settings** dialog box will be displayed.

- Specify the size of mesh element as **7** mm in the **Element Size** edit box and click on the **Generate Mesh** button. The mesh will be generated.
- Click on the **OK** button from the dialog box.
- Click on the **Run** tool from the **Solve** panel of **Autodesk Inventor Nastran** tab in the **Ribbon**. Once the analysis is complete, **Autodesk Inventor Nastran** information box will be displayed.
- Click on the **OK** button from the information box. The results of analysis will be displayed; refer to Figure-9.

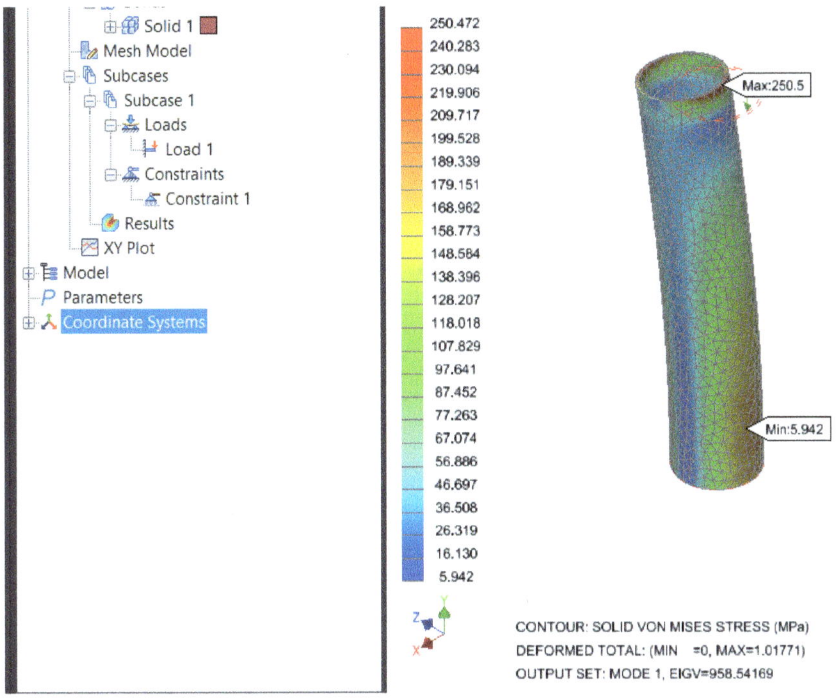

Figure-9. Result of buckling analysis

PRACTICE

Perform buckling analysis on the models as shown in Figure-10 and compare the results. The tube is fixed at the bottom and force of **400** N is applied on the top face.

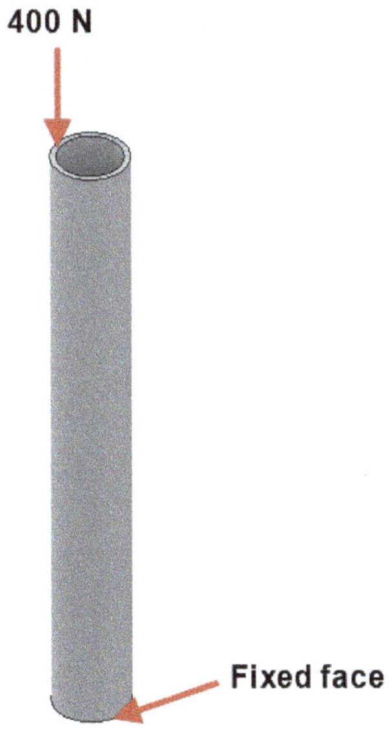

Figure-10. Model for buckling analyses

For Student Notes

Chapter 6

Transient Response Analysis

Topics Covered

The major topics covered in this chapter are:

- ***Introduction.***
- ***Performing Direct Transient Response Analysis***
- ***Defining Damping Parameters***
- ***Defining Dynamic Setup Parameters***
- ***Performing Modal Transient Response Analysis***
- ***Performing Nonlinear Transient Response Analysis***

INTRODUCTION

Transient response analysis is a method of computing forced dynamic response. This analysis is used to determine the behavior of a structure subjected to time-varying excitation. During the transient analysis, loads applied to the structure are known at each instant in time. Loads can be in the form of applied forces and enforced motions. The results obtained from a transient response analysis are in the form of displacements, velocities, and accelerations of grid points, or forces and stresses in elements at each output time step.

Depending upon the structure and the nature of the loading, there are three different transient response analyses:

Direct Transient Response Analysis: This analysis calculates the response of a system to a load over time. The load applied to the system can vary over time or simply be an initial condition that is allowed to evolve over time. This method may be more efficient for models where high-frequency excitation require the extraction of a large number of modes. Also, if structural damping is used, the direct method should be used.

Modal Transient Response Analysis: This analysis is an alternate technique available for dynamics that utilizes the mode shapes of the structure, reduces the solution degrees of freedom, and can significantly impact the run time. This approach replaces the physical degrees of freedom with a reduced number of modal degrees of freedom. Fewer degrees of freedom mean a faster solution. This can be a big time saver for transient models with a large number of time steps. Because modal transient response analysis uses the mode shapes of a structure, this analysis is a natural extension of normal modes analysis.

Nonlinear Transient Response Analysis: This analysis is performed when effect of inertia, damping and transient loading are significant. Also if your model is in unstable conditions or can go under buckling deformation then this analysis can give more insight of failure. An important element to having a stable nonlinear transient (NLT) solution is to provide damping in the model. Note that while using damping in a NLT solution for models where the velocity/inertia is the main driver of the analysis such as in an impact solution, damping can have a significant effect on the acceleration/velocity/displacement of the model. This is because the solver cannot make a distinction between rigid body motion/velocity, and flexible motion/velocity, so the damping is applied to any part of the structure that has a velocity. For impact analysis it is recommended to use no damping or a small "stability" damping value.

The procedures to perform these analyses are discussed next.

PERFORMING DIRECT TRANSIENT RESPONSE

The direct transient response analysis is used for structural objects under high frequency excitation. The procedure to perform direct transient response analysis is given next.

- Open the part on which you want to perform analysis and start Autodesk Inventor Nastran environment. Linear Static analysis will become active automatically.
- Click on the **Edit** tool from the **Analysis** panel of **Autodesk Inventor Nastran** tab in the **Ribbon**. The **Analysis** dialog box will be displayed.
- Select the **Direct Transient Response** option from the **Type** drop-down. Set the other options in the **Analysis** dialog box as discussed earlier and click on the **OK** button. The options in the **Model Tree** will be displayed as shown in Figure-1.

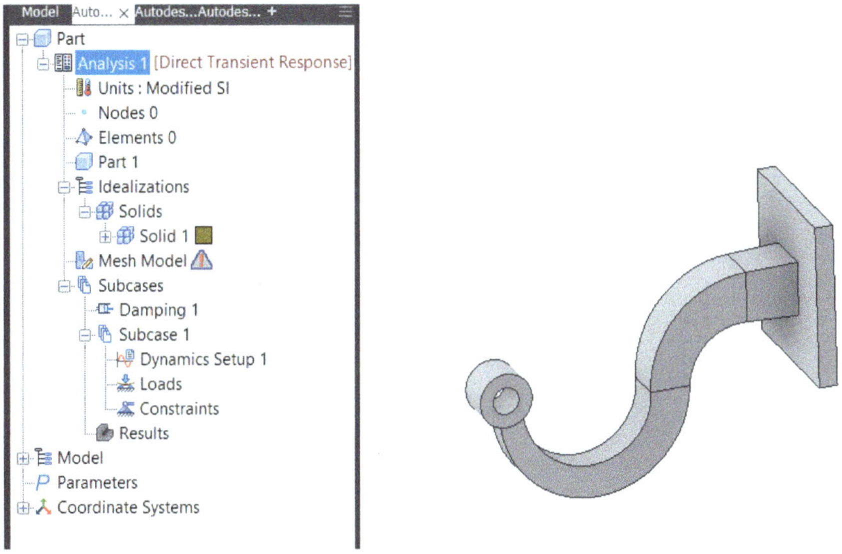

Figure-1. Options for direct transient response analysis

Setting Damping Values

- Double-click on **Damping 1** option from the **Subcases** node. The **Damping** dialog box will be displayed; refer to Figure-2.

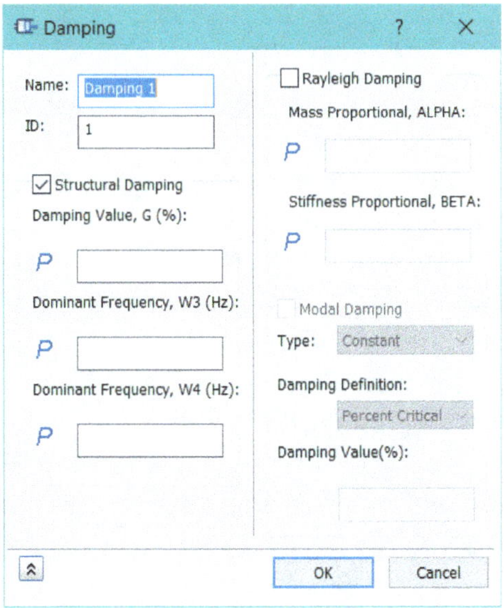

Figure-2. Damping dialog box

- Specify desired name and id for damping in **Name** and **ID** edit boxes.
- Specify the desired value of global damping in the **Damping Value, G(%)** edit box. The value specified here is in percentage.
- Specify the desired value of frequency in the **Dominant Frequency, W3(Hz)** edit box at which elemental damping will be converted to equivalent viscous damping. It is a global structural damping setting.
- Specify the desired value of frequency in the **Dominant Frequency, W4(Hz)** edit box at which elemental damping will be converted to equivalent viscous damping. It is a material structural damping setting.
- Select the **Rayleigh Damping** check box to define mass and stiffness components of rayleigh damping. The edit boxes below the check box will become active.
- Specify the desired value of mass proportion and stiffness proportion in respective edit boxes.
- Set the desired values and click on the **OK** button.

Dynamic Setup Parameters

- Double-click on **Dynamic Setup 1** option from the **Subcase 1** node. The **Dynamics Setup** dialog box will be displayed; refer to Figure-3.

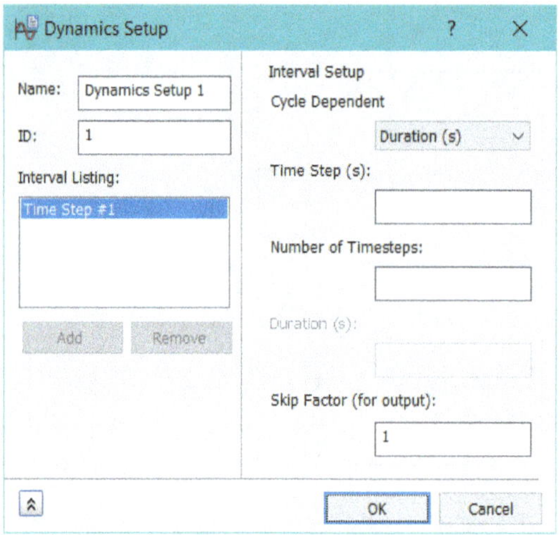

Figure-3. Dynamics Setup dialog box

- Set the desired name and id in **Name** and **ID** edit boxes.
- Select the desired option from the **Cycle** Dependent drop-down. Select the **Time Step(s)** option from the drop-down if you want to specify number of time steps and total duration for which you want to run the analysis. Select the **Number of Timesteps** option from the drop-down if you want to specify duration of each time steps and total duration for which you want to run the analysis. Select the **Duration(s)** option from the drop-down if you want to duration of each time step and total number of time steps.
- Specify the desired skip factor in the **Skip Factor (for output)** edit box and set the number for time step to be skipped.
- After setting desired parameters, click on the **OK** button.

- Specify the desired idealization, load, and constraint for the analysis.
- After setting the desired parameters and generating the mesh, click on the **Run** tool from the **Solve** panel.
- Once the analysis is complete, an information box will be displayed. Click on the **OK** button from the information box. The results of analysis will be displayed; refer to Figure-4.

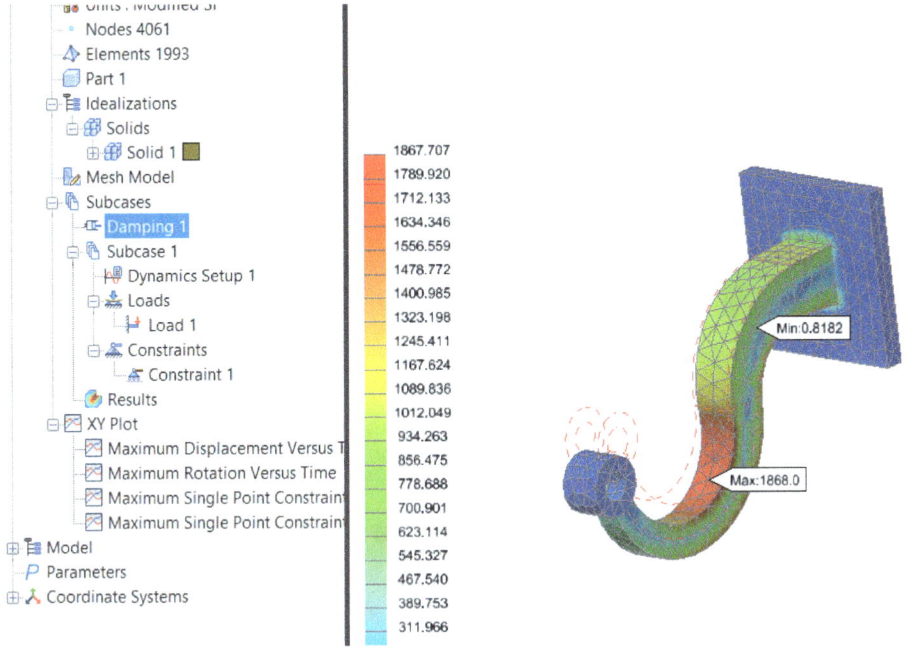

Figure-4. Result of direct transient analysis

- Double-click on desired result in **XY Plot** node to check plots.

PERFORMING MODAL TRANSIENT RESPONSE ANALYSIS

The modal transient response analysis is performed when you want to use modal degrees of freedom in place of physical degrees of freedom for solving the analysis. The procedure to perform analysis is given next.

- Open the model on which you want to perform modal transient response analysis and start **Autodesk Inventor Nastran** environment. The linear static analysis will start automatically.
- Click on the **Edit** tool from the **Analysis** panel of **Autodesk Inventor Nastran** tab in the **Ribbon**. The **Analysis** dialog box will be displayed.
- Select the **Modal Transient Response** option from the **Type** drop-down and set the desired parameters.
- Click on the **OK** button from the dialog box. The options will be displayed as shown in Figure-5.

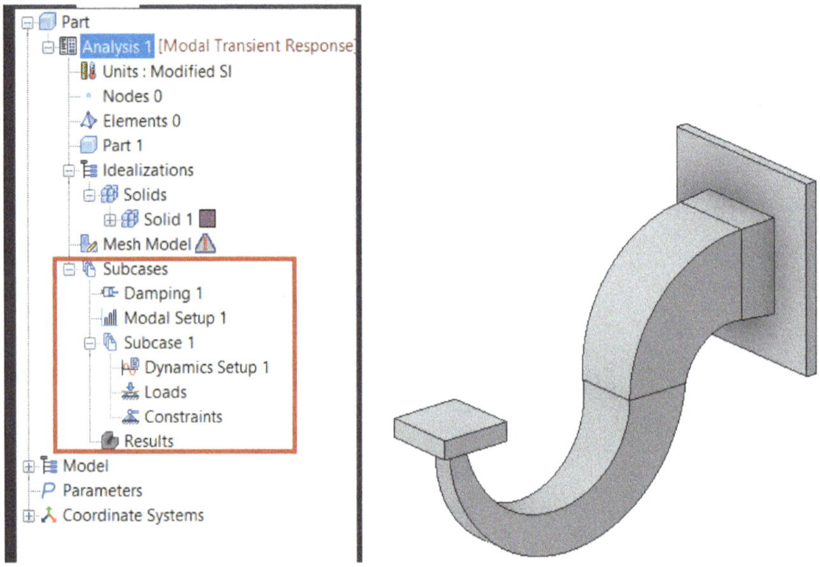

Figure-5. Options for modal transient response analysis

- Apart from direct transient response analysis options, modal setup options will also be displayed. Double-click on **Modal Setup 1** option from the **Subcases** node. The **Modal Setup** dialog box will be displayed.
- Set the desired parameters as discussed earlier and click on the **OK** button.
- Similarly, set the modal damping parameters in the **Modal Damping** area of **Damping** dialog box.
- Apply material, idealization, load, and constraint as desired and click on the **Run** button. Once the analysis is complete, click on the **OK** button from the **Autodesk Inventor Nastran** information box. The analysis results will be displayed; refer to Figure-6.

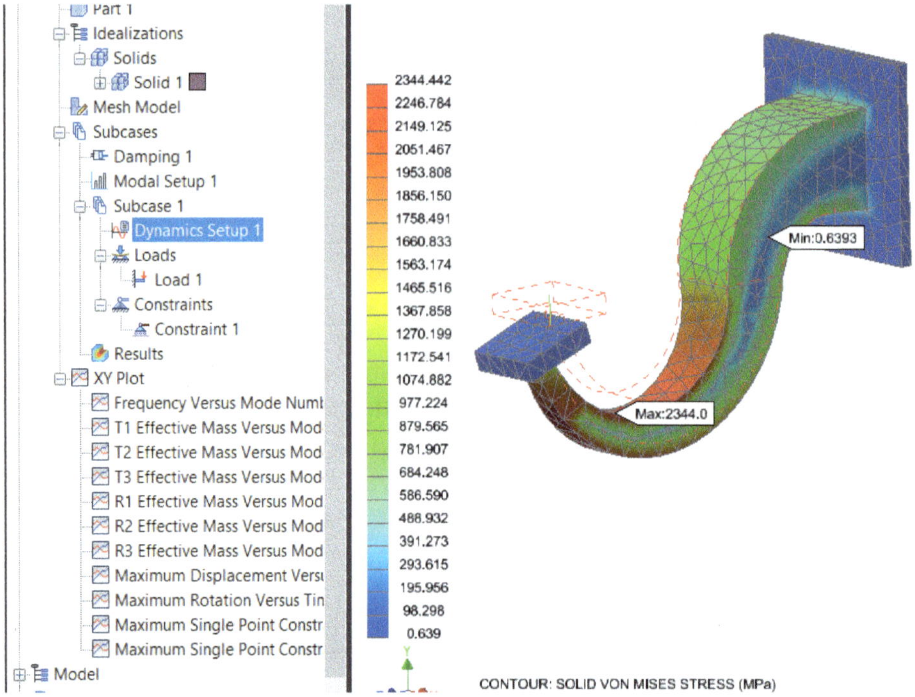

Figure-6. Modal transient response analysis results

- Double-click on desired plot to check the results.

PERFORMING NONLINEAR TRANSIENT RESPONSE ANALYSIS

The nonlinear transient response analysis is performed when damping, vibrational excitation, and large deformation is involved in the analysis study. The procedure to perform the analysis is given next.

- Open the part on which you want to perform nonlinear transient response analysis and start **Autodesk Inventor Nastran** environment.
- Click on the **New** tool from the **Analysis** panel of **Autodesk Inventor Nastran** tab in the **Ribbon**. The **Analysis** dialog box will be displayed.
- Select **Nonlinear Transient Response** option from the **Type** drop-down of the dialog box.
- Click on the **Options** tab in the dialog box and select the **On** option from the **Large Displacements** drop-down if you are expecting large deformation.
- Click on the **OK** button from the dialog box. The options to set parameters for nonlinear transient analysis will be displayed in the **Model Tree**; refer to Figure-7.

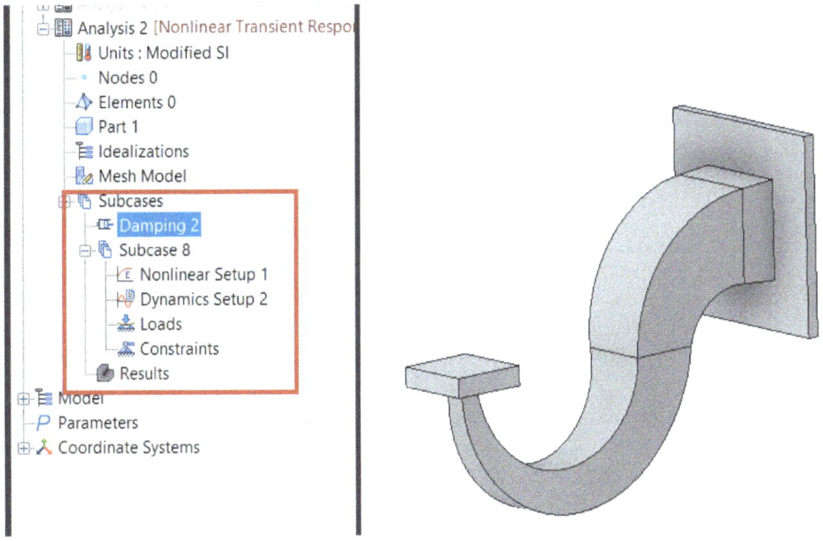

Figure-7. Options for nonlinear transient response analysis

- Set the desired parameters for damping, nonlinear setup, and dynamic setup.
- Set the load and constraints as required.
- Specify material and set the idealization.
- Click on the **Run** button from the **Solve** panel of **Autodesk Inventor Nastran** tab in the **Ribbon**. Once the analysis is complete, click on the **OK** button from the **Autodesk Inventor Nastran** information box. The analysis results will be displayed; refer to Figure-8.

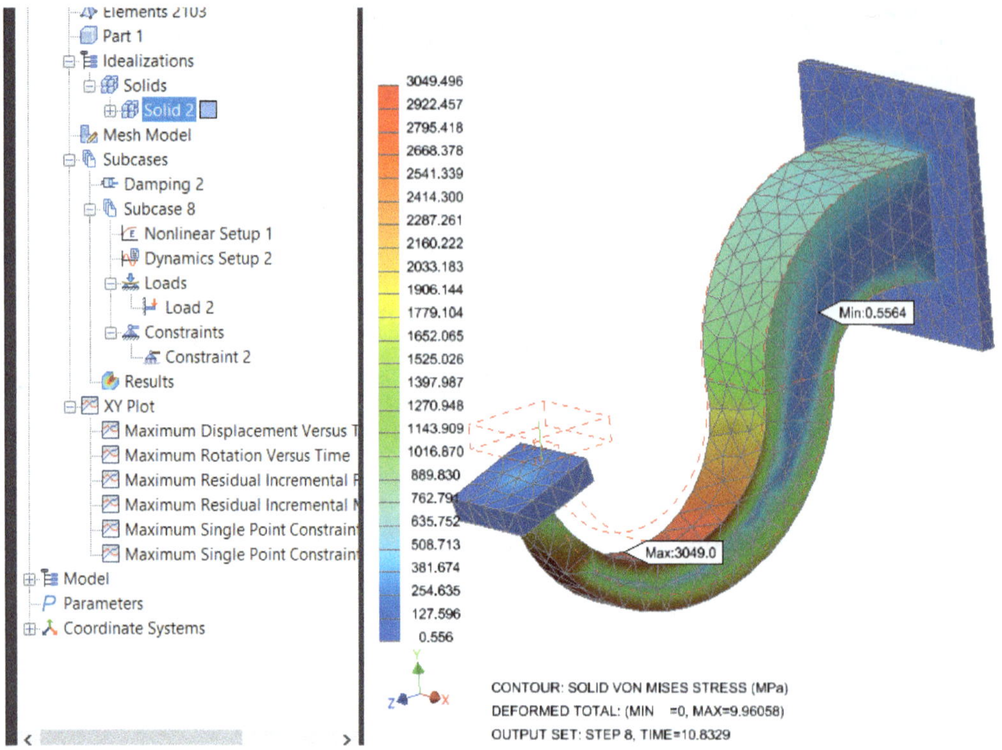

Figure-8. Result of nonlinear transient response analysis

You can also check results at different time steps. Save the results and exit the environment.

PRACTICE

Perform nonlinear transient response analysis on the leaf spring as shown in Figure-9.

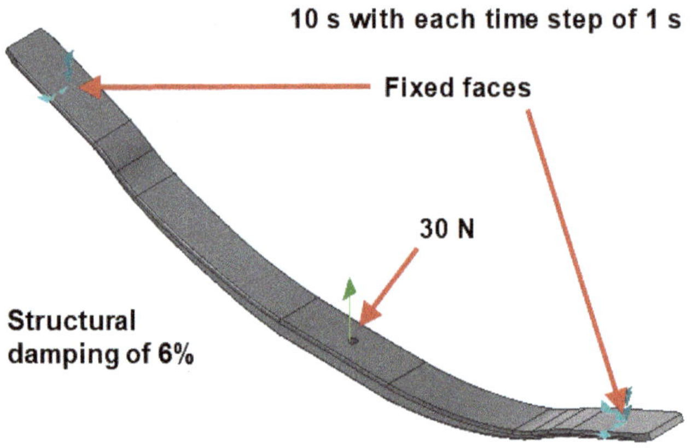

Figure-9. Model for nonlinear transient response analysis

Chapter 7

Impact Analysis

Topics Covered

The major topics covered in this chapter are:

- ***Introduction.***
- ***Performing Impact Analysis***
- ***Creating Sketch line for Impact direction***
- ***Defining Impact Parameter***

INTRODUCTION

The Impact analysis is performed to check drop-test and projectile impact studies. The input parameters used in this analysis are direction of travel, initial velocity, and acceleration of part.

PERFORMING IMPACT ANALYSIS

The procedure to perform impact analysis is given next.

- Open the model on which you want to perform analysis and start **Autodesk Inventor Nastran** environment; refer to Figure-1.

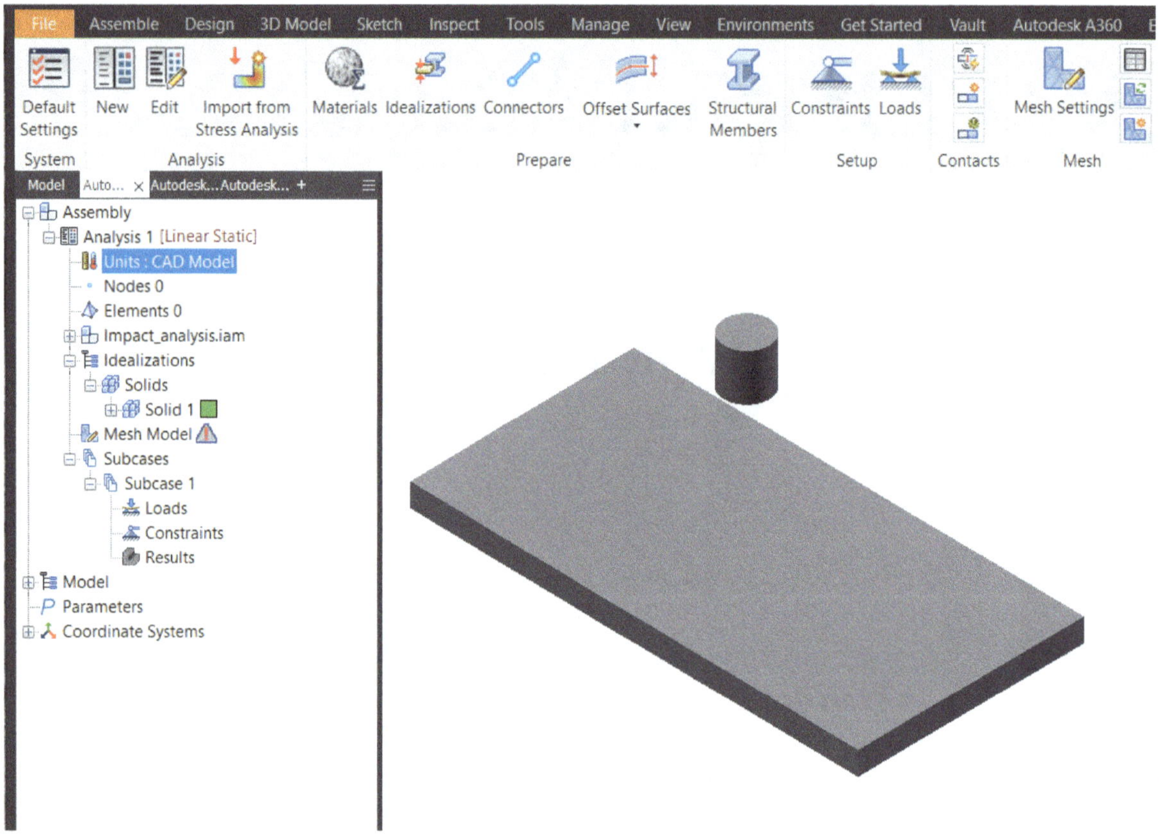

Figure-1. Assembly model for impact analysis

- Click on the **Start 2D Sketch** tool from the **Sketch** panel of **Sketch** tab in the **Ribbon**. You will be asked to select sketching plane.
- Click on the **Model** tab in the **Model Tree** and select YZ plane as shown in Figure-2. The sketching tools will be displayed.
- Create the line with end points coincident to model; refer to Figure-3.

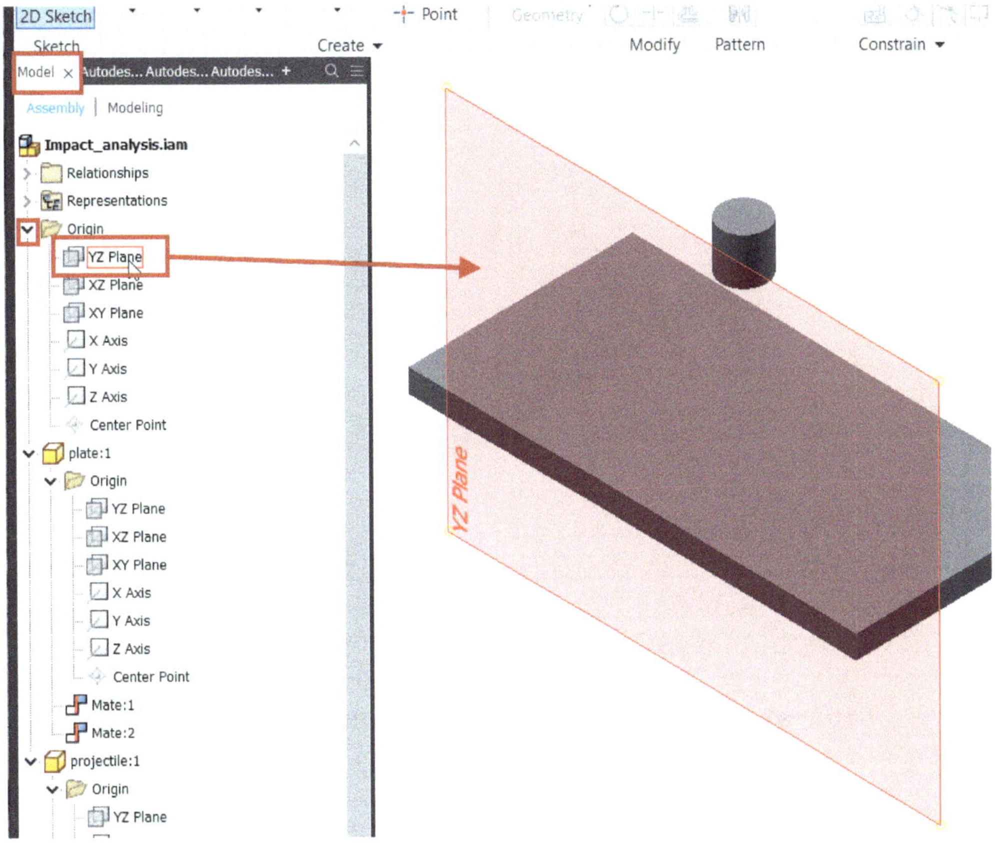

Figure-2. Selecting plane for sketch

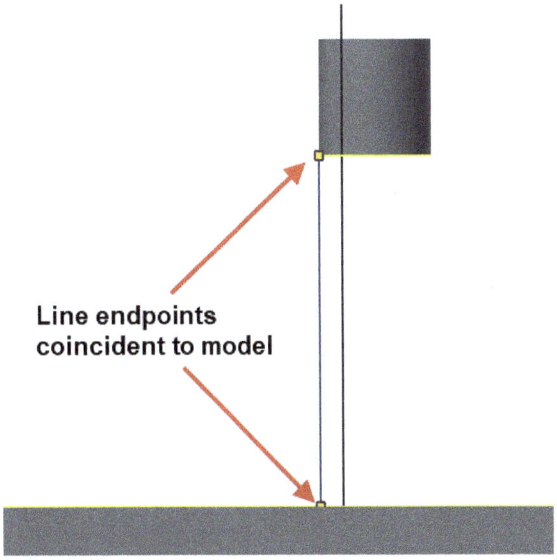

Figure-3. Creating line

- Click on the **Finish Sketch** tool from the **Exit** panel of **Sketch** tab in the **Ribbon**. The model will be displayed as shown in Figure-4.

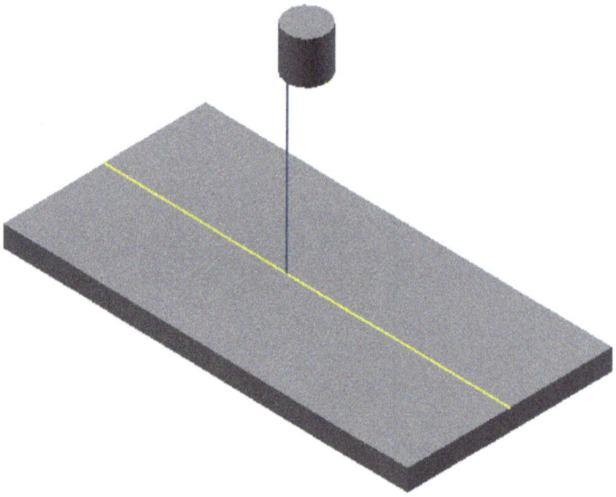

Figure-4. Model with sketch

- Click on the **Edit** button from the **Analysis** panel of **Autodesk Inventor Nastran** tab in the **Ribbon**. The **Analysis** dialog box will be displayed.
- Select the **Impact Analysis** option from the **Type** drop-down.
- Select the **On** option from the **Large Displacements** drop-down and select the **Separation** option from the **Contact Type** drop-down in the **Options** tab.
- Set the other options as required and click on the **OK** button.

Defining Impact Parameters

- Double-click on **Impact Setup** option from the **Subcases** node. The **Impact Analysis** dialog box will be displayed; refer to Figure-5.

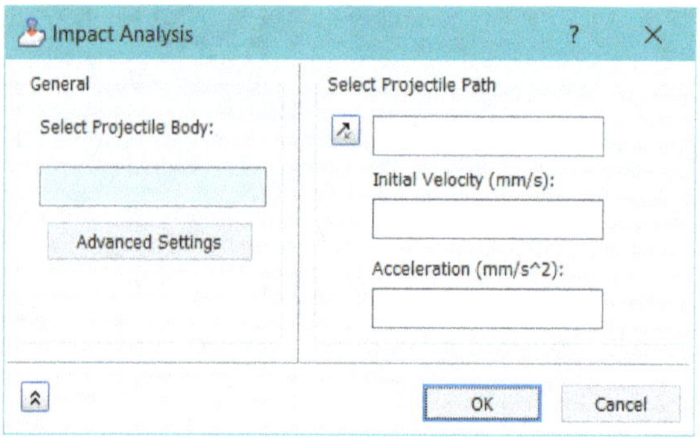

Figure-5. Impact Analysis dialog box

- Select the part that you want to throw in impact analysis.
- Click in the **Projectile Translation Vector** selection box of **Select Projectile Path** area and select the sketch line earlier created to define direction in which projectile part will move.
- Specify the desired value for initial velocity and acceleration in respective edit boxes of the dialog box; refer to Figure-6.

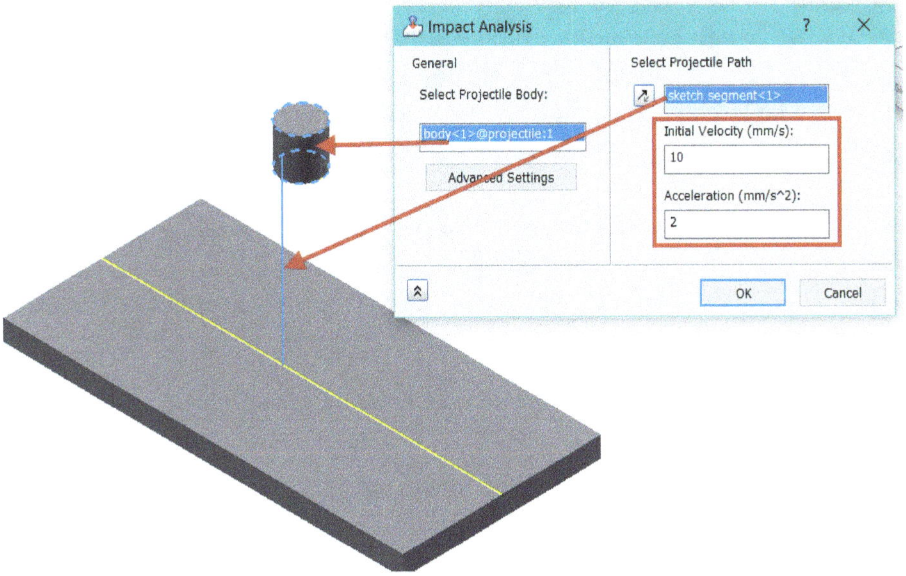

Figure-6. Setting parameters for impact analysis

- Click on **Advanced Settings** button from the dialog box. The **Impact Event Parameters** dialog box will be displayed; refer to Figure-7.

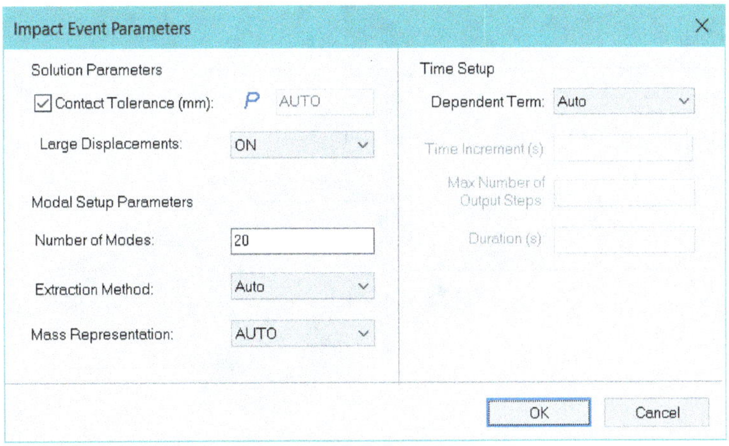

Figure-7. Impact Event Parameters dialog box

- Specify the desired number of modes to be generated for each time step.
- Select the desired option from the **Dependent Term** drop-down of **Time Setup** area and define the values of related parameters; refer to Figure-8.

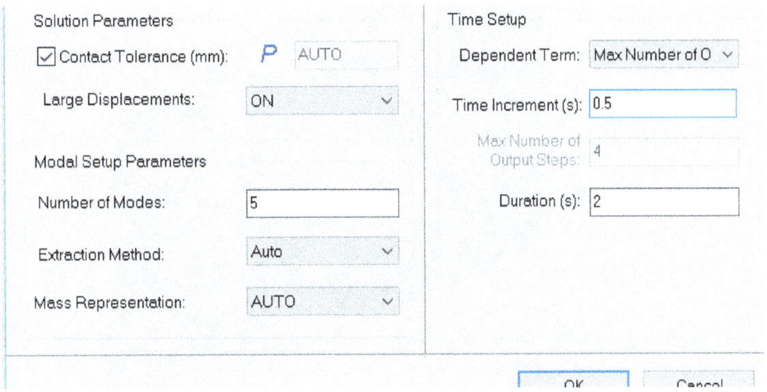

Figure-8. Setting parameters

- After setting the desired values, click on the **OK** button.
- Click on the **Materials** tool from the **Prepare** panel of **Autodesk Inventor Nastran** tab in the **Ribbon**. The **Material** dialog box will be displayed.
- Click on the **Select Material** button from the dialog box and select **Alloy Steel** material in **Inventor Material Library** node of the **Material DB** dialog box.
- Click on the **OK** button from the **Material DB** dialog box and then click on the **OK** button from the **Material** dialog box.
- Double-click on **Solid 1** option from the **Idealizations** node. The Idealizations dialog box will be displayed.
- Select **Alloy Steel** option from **Material** drop-down and click on the **OK** button.
- After setting desired parameters, click on the **Run** tool from the **Solve** panel of **Autodesk Inventor Nastran** tab in the **Ribbon**. The result of analysis will be displayed at different time steps; refer to Figure-9.

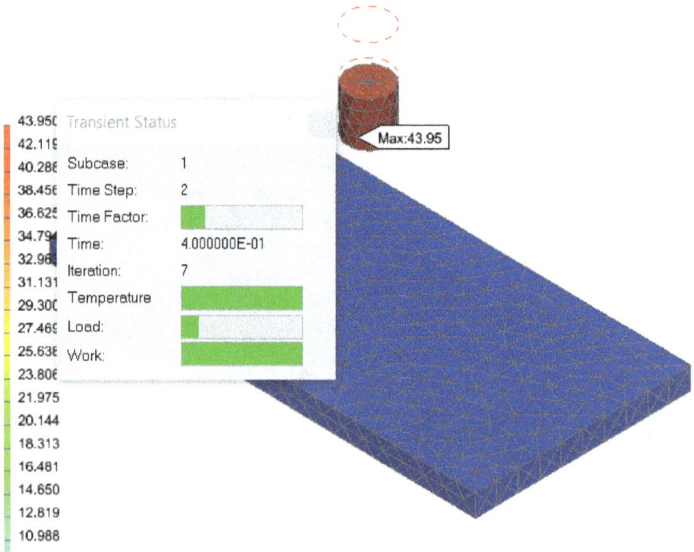

Figure-9. Result of impact analysis

PRACTICE

Perform impact analysis on the model shown in Figure-10.

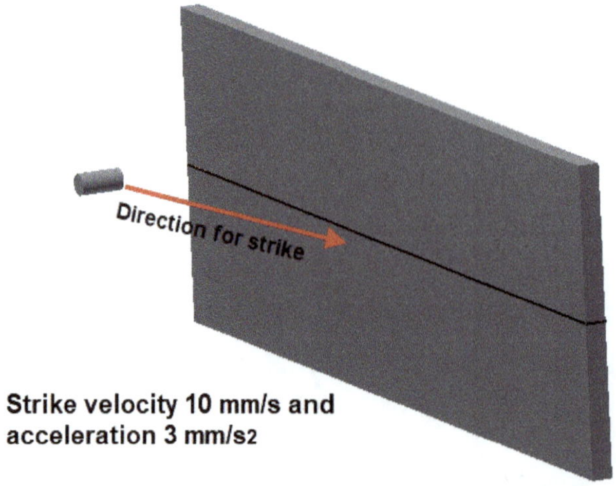

Figure-10. Model for Practice

Chapter 8

Frequency Response Analysis

Topics Covered

The major topics covered in this chapter are:

- ***Introduction***
- ***Performing Direct Frequency Response Analysis***
- ***Performing Modal Frequency Response Analysis***
- ***Defining Damping, Modal Setup, and Dynamic Setup Parameters***

INTRODUCTION

The frequency response analysis is used to compute structural response of model under steady state oscillation. There are two types of frequency response analyses available in the Autodesk Inventor Nastran; Direct Frequency Response and Modal Frequency Response.

Direct Frequency Response analysis is performed when structural response is to be found at discrete frequencies. This analysis is useful when large number of modes are to be extracted at high frequency excitation.

Modal Frequency Response analysis is performed when large size model is to be checked under specified mode excitations. This analysis uses the mode shapes of the structure to uncouple the equations of motion (when no damping or only modal damping is used) and, depending on the number of modes computed and retained, the problem size is reduced. Both of these factors tend to make modal frequency response analysis computationally more efficient than direct frequency response analysis.

The procedures to perform these analyses are given next.

PERFORMING DIRECT FREQUENCY RESPONSE ANALYSIS

The procedure to perform direct frequency response analysis is given next.

- Open the part on which you want to perform direct frequency response analysis and start Autodesk Inventor Nastran environment.
- Click on the **Edit** button from the **Analysis** panel of **Autodesk Inventor Nastran** tab in the **Ribbon**. The **Analysis** dialog box will be displayed.
- Select the **Direct Frequency Response** option from the **Type** drop-down and set the desired parameters.
- Click on the **OK** button from the dialog box. The options will be displayed as shown in Figure-1.
- Double-click on damping option in subcases node and set the damping parameters as discussed earlier.

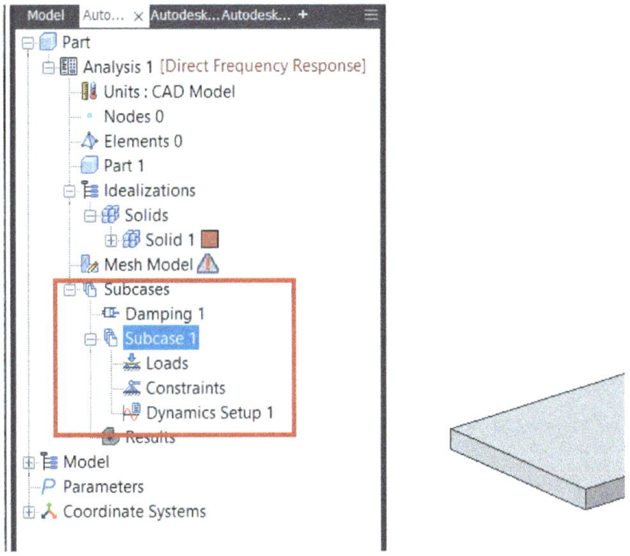

Figure-1. Subcase options for direct frequency response

- Double-click on **Dynamics Setup 1** option from the **Subcase 1** node. The **Dynamics Setup** dialog box will be displayed; refer to Figure-2.

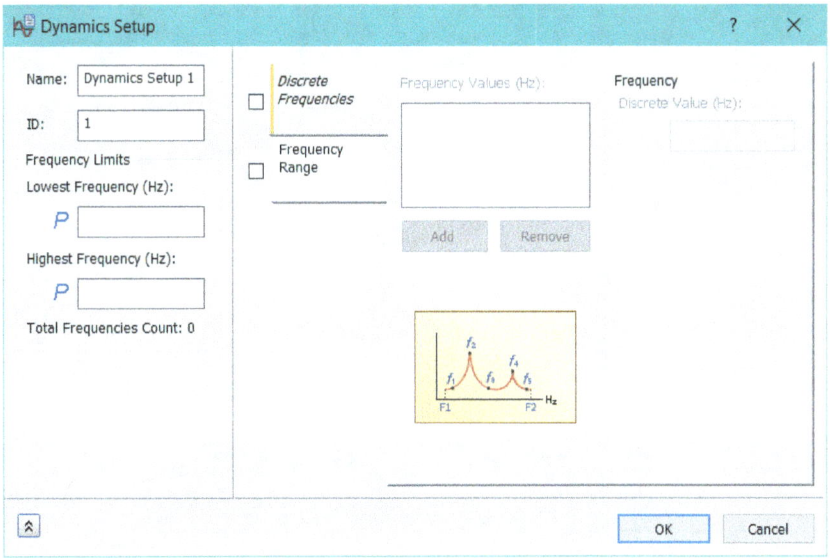

Figure-2. Dynamics Setup dialog box

- Set the desired upper and lower limits of frequency range in the respective edit boxes at the left in the dialog box.
- Select the **Discrete Frequencies** check box if you want to specify values of frequencies explicitly for excitation of model. The options in the right area will become active. Specify desired value in the **Discrete Value (Hz)** edit box and click on the **Add** button to add it in the list of frequencies. If you want to remove any frequency from the list then select it from the list and click on the **Remove** button.
- Select the **Frequency Range** check box to define range of frequencies to be used for excitation of system. The options in the right will become active. Specify the number of frequencies to be generated in the specified range using the **Number of Points in the Range** edit box. The value in **Frequency Increment** edit box will be generated automatically.

- After setting desired values, click on the **OK** button.
- Set the idealization, constraint, and load parameters as discussed earlier and click on the **Run** button. The analysis result will be displayed; refer to Figure-3.

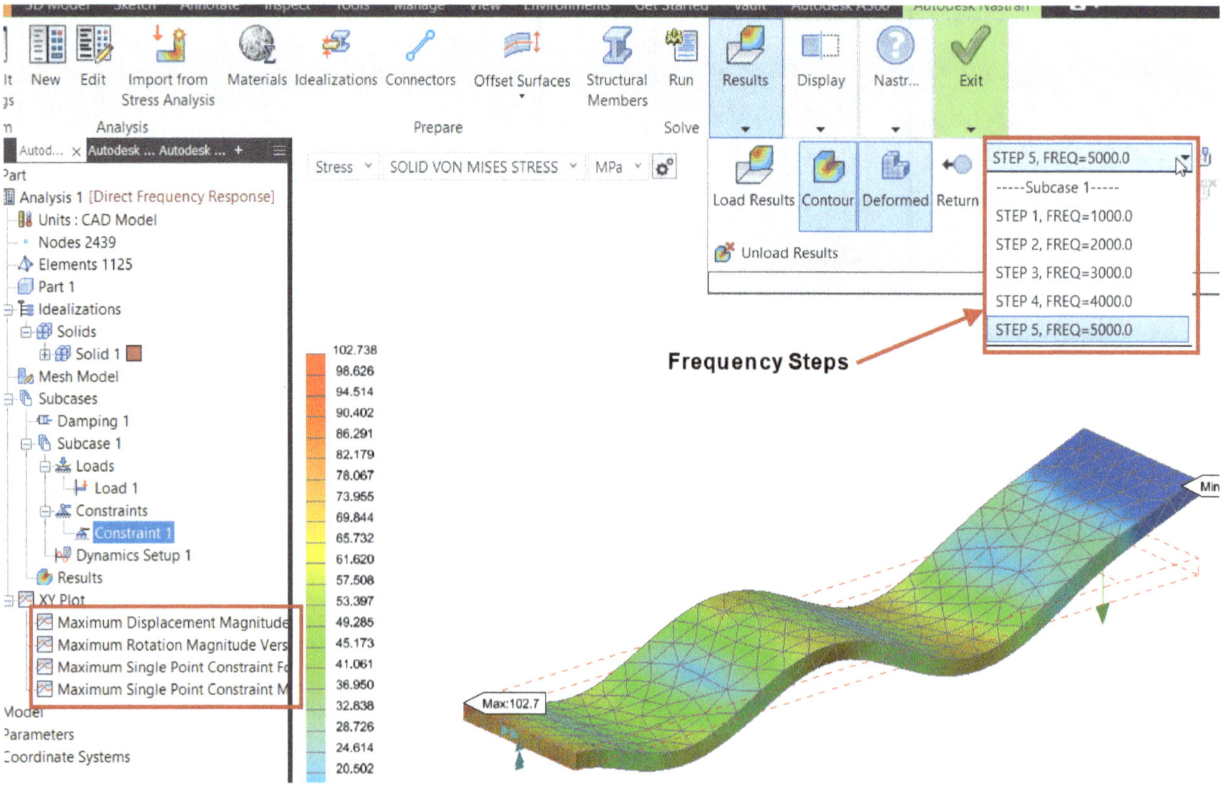

Figure-3. Result of direct frequency analysis

- Select the desired frequency step and double-click on result nodes to check the results.

PERFORMING MODAL FREQUENCY RESPONSE ANALYSIS

The procedure to perform modal frequency response analysis is given next.

- Open the part on which you want to perform modal frequency response analysis and start **Autodesk Inventor Nastran** environment.
- Click on the **New** button from the **Analysis** panel of **Autodesk Inventor Nastran** tab in the **Ribbon**. The **Analysis** dialog box will be displayed.
- Select the **Modal Frequency Response** option from the **Type** drop-down and set the other options as required.
- Click on the **OK** button from the dialog box. The options related to modal frequency response analysis will be displayed; refer to Figure-4.

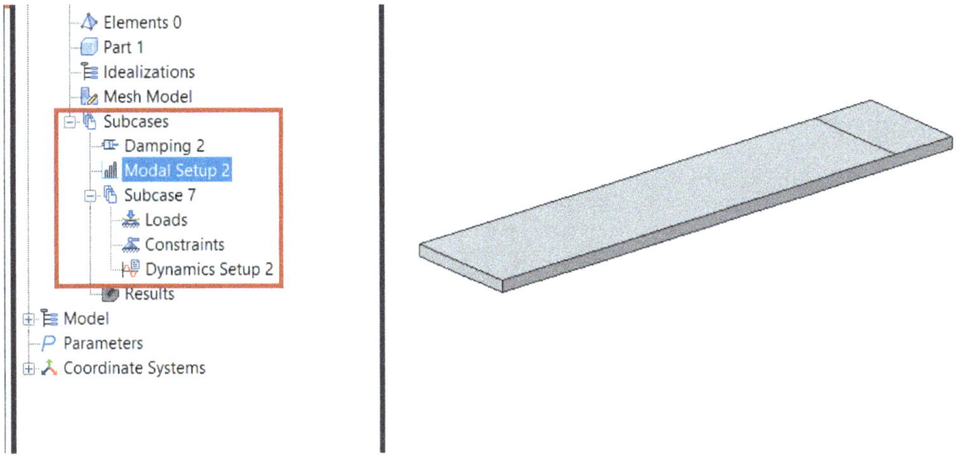

Figure-4. Subcase options for modal frequency response

- Double-click on **Damping** option from the **Subcases** node and set the desired structural, modal, and rayleigh damping parameters.
- Double-click on **Modal Setup** option in **Subcases** node and set the desired parameters; refer to Figure-5.

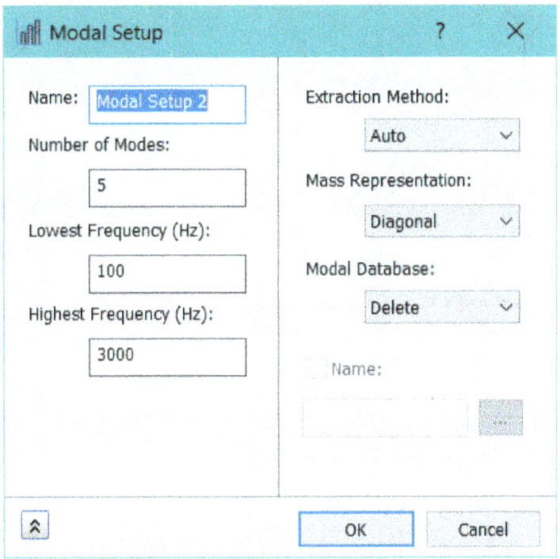

Figure-5. Modal setup options

- Double-click on **Dynamics Setup** option from the **Subcase** node. The **Dynamics Setup** dialog box will be displayed as shown in Figure-6.

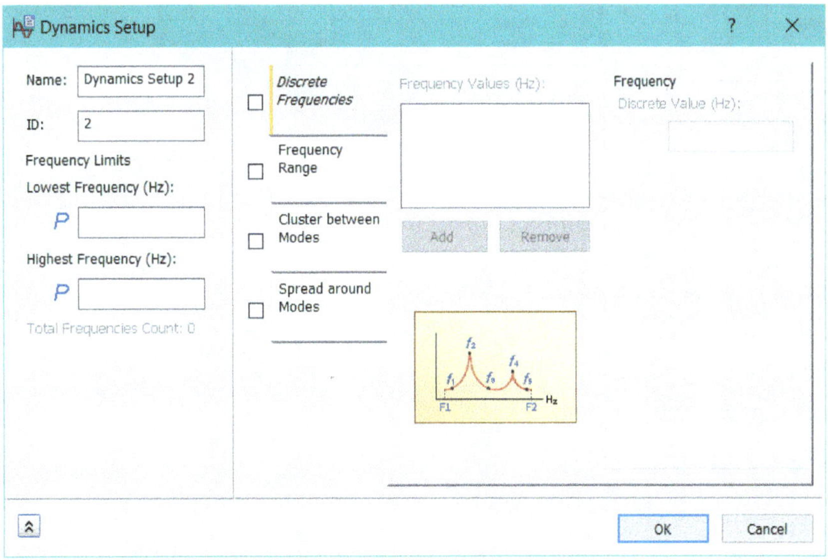

Figure-6. Dynamics Setup dialog box

- Set the desired discrete frequencies and frequency range in the dialog box using **Discrete Frequencies** and **Frequency Range** check boxes.
- Select the **Cluster between Modes** check box to define different frequency range clusters and set the desired parameters; refer to Figure-7.

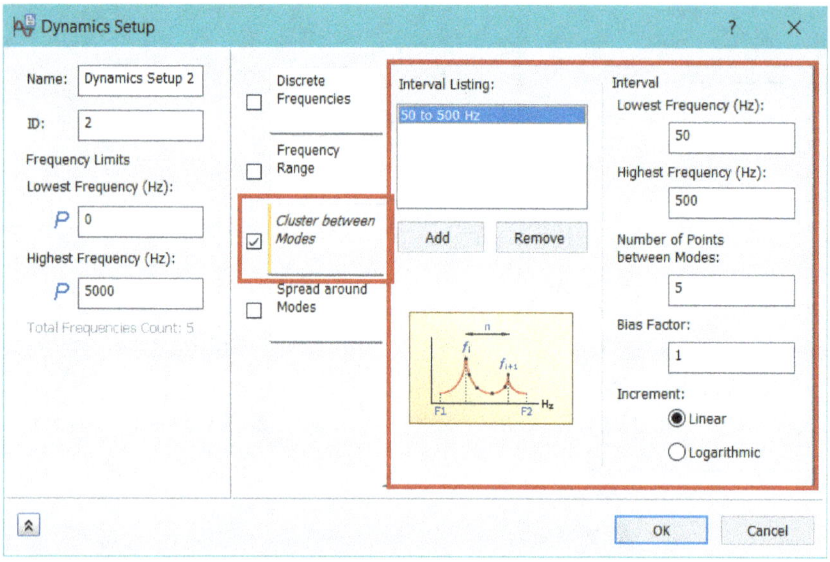

Figure-7. Cluster between Modes parameters

- Select the **Spread around Modes** check box to define parameters related to frequency distribution and set the desired parameters.
- Set the other parameters as discussed earlier and click on the **Run** button from the **Solve** panel. The result of analysis will be displayed; refer to Figure-8.

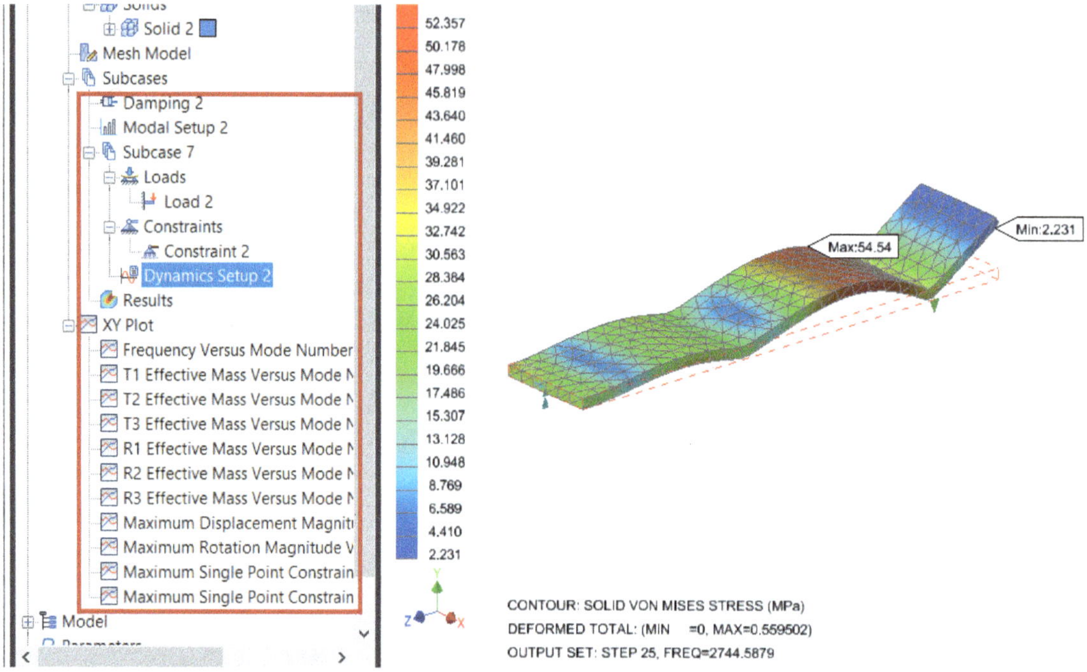

Figure-8. Result of modal frequency analysis

Select the desired frequency and double-click on desired result node to check the results of analysis.

Chapter 9

Random Response Analysis

Topics Covered

The major topics covered in this chapter are:

- ***Introduction.***
- ***Performing Random Response Analysis***
- ***Defining Dynamic Setup Parameters***
- ***Setting PSD Output Control***
- ***Defining Summation Method for Structural Response***
- ***Performing Shock/Response Spectrum Analysis***
- ***Defining Shock Spectrum Curve***

INTRODUCTION

The Random Response analysis is performed when the model is under vibrational conditions such as earthquake ground motion, ocean wave heights and frequencies, wind pressure fluctuations on aircraft and tall buildings, and acoustic excitation due to rocket and jet engine noise. In such vibrational conditions, exact values of frequencies of excitation are not known but you know the mean value, standard deviation, and probability of exceeding a certain value.

PERFORMING RANDOM RESPONSE ANALYSIS

The procedure to perform random response analysis is given next.

- Open the part on which you want to perform random response analysis and start **Autodesk Inventor Nastran**. Linear static analysis will activate automatically.
- Click on the **Edit** button from the **Analysis** panel of **Autodesk Inventor Nastran** tab in the **Ribbon**. The **Analysis** dialog box will be displayed.
- Select the **Random Response** option from the **Type** drop-down. The options related to random response will be displayed.
- Click on the **PSD Output Control** button from the **Dynamic Options** area to define output control parameters for nodes and elements. The **Output Control** dialog box will be displayed; refer to Figure-1.

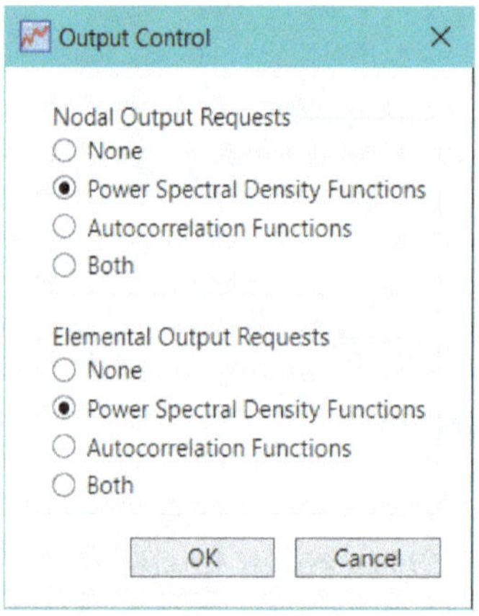

Figure-1. Output Control dialog box

- Select the desired radio buttons from the dialog box and click on the **OK** button.
- Specify desired parameters in the **Analysis** dialog box and click on the **OK** button.
- Set the desired damping, modal setup, loads, and constraints as discussed earlier.
- Double-click on **Dynamic Setup** option under **Subcase 1** node. The **Dynamics Setup** dialog box will be displayed.
- Click on the **New Table** button in **Random Analysis Options** area of the dialog box. The **Table Data** dialog box will be displayed; refer to Figure-2.

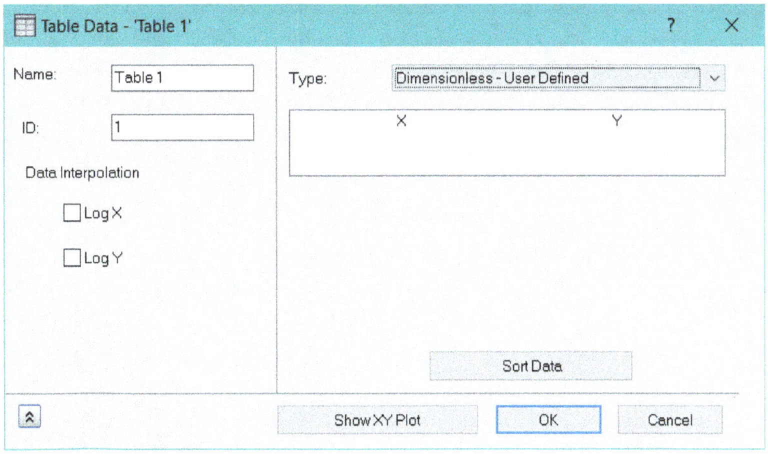

Figure-2. Table Data dialog box

- Select the **PSD (Acceleration) vs. Frequency** option from the **Type** drop-down if you want to specify acceleration for each specified frequency value. Select the **PSD (Force) vs. Frequency** option from the **Type** drop-down if you want to specify force for each specified frequency value; refer to Figure-3.

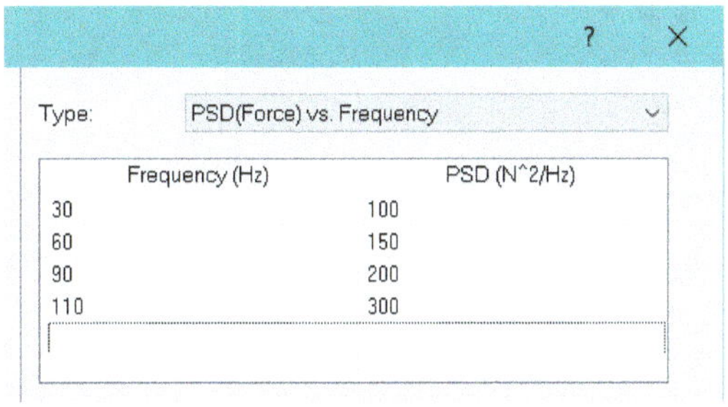

Figure-3. Specifying force vs frequency values

- After setting the desired values, click on the **OK** button from the **Table Data** dialog box.
- Set the other parameters in **Dynamics Setup** dialog box as discussed earlier and click on the **OK** button from the dialog box.
- Click on the **Run** button from the **Solve** panel and perform the analysis. Once the analysis is complete, the results will be displayed; refer to Figure-4.

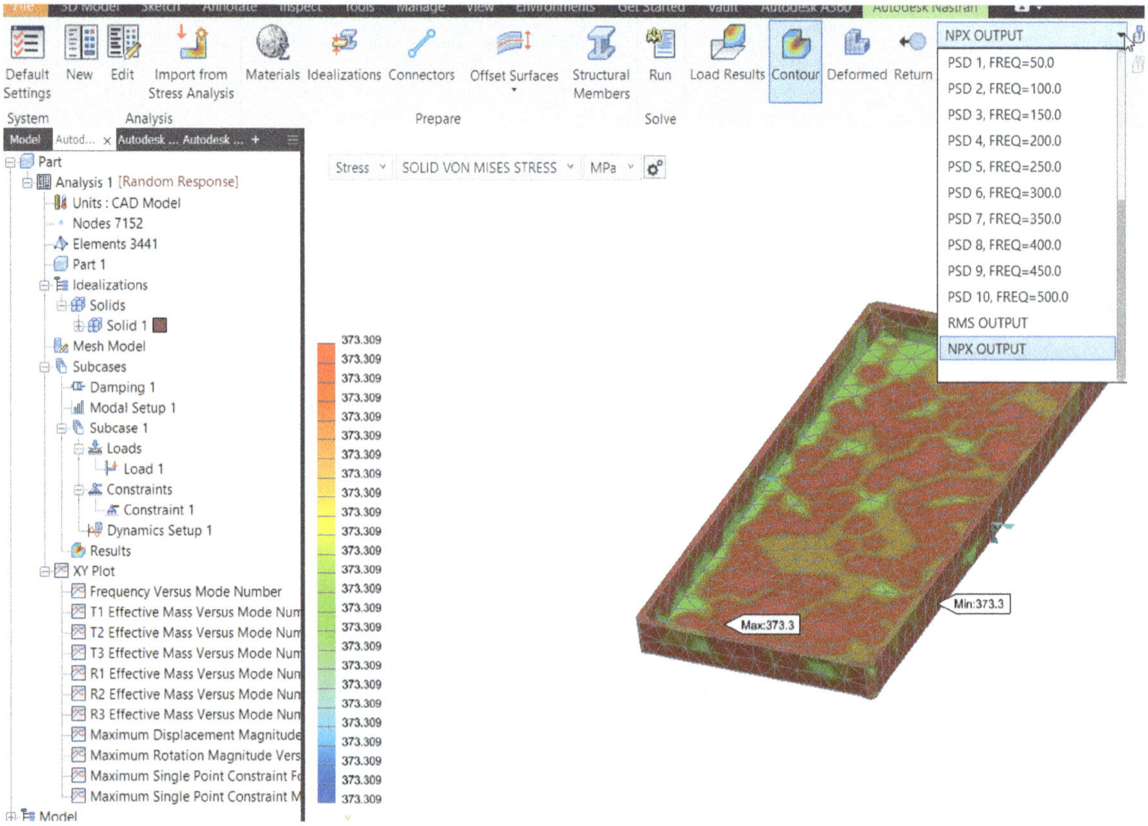

Figure-4. Result of random response analysis

- Select the desired frequency from **Results** drop-down in **Results** panel to check the results at that specific frequency.

PERFORMING SHOCK/RESPONSE SPECTRUM ANALYSIS

The Shock/Response Spectrum analysis is performed when there is a shock load on the model and you want to check the results in the form of a stress frequency band. The procedure to perform this analysis is given next.

- Open the model on which you want to perform shock analysis and start Autodesk Inventor Nastran environment.
- Click on the **Edit** button from the **Analysis** panel of **Autodesk Inventor Nastran** tab in the **Ribbon**. The **Analysis** dialog box will be displayed.
- Select the **Shock/Response Spectrum** option from the **Type** drop-down of the dialog box.
- Click on the **Options** tab to specify parameters for spectrum. The options will be displayed as shown in Figure-5.

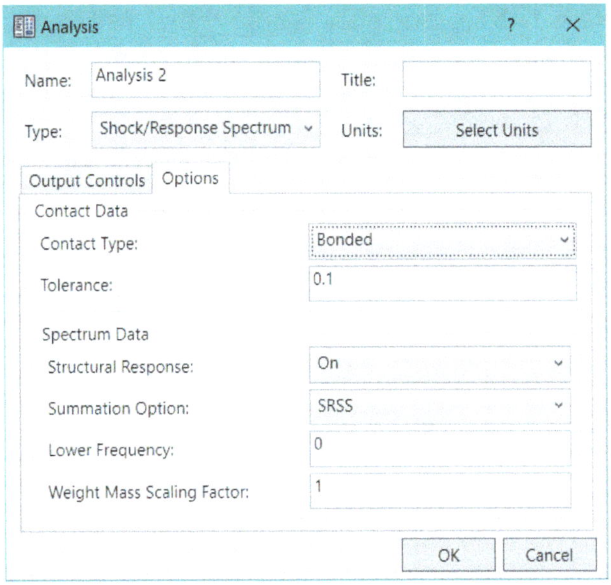

Figure-5. Response spectrum options

- Select the **On** option from the **Structural Response** drop-down if you want to include structural response of shock as well along with modal response.
- Select the desired option from the **Summation Option** drop-down to define how modal results will be combined. Select the **ABS** option if you want to add absolute values of modal results in the analysis result. Select the **SRSS** option (Square Root of Sum of Squares) if you want to add squares of modal results and square root the final value. Select the **NRL** option (Naval Research Laboratory) if you want to combine ABS and SRSS methods to get find result. Select the **CQC** option (Complete Quadratic Combination) from the drop-down if you want to use the quadratic combination formula.
- Specify the desired value of lower frequency and weight mass scaling factor in respective edit boxes for the analysis.
- After setting the desired parameters, click on the **OK** button from the dialog box.
- Double-click on **Dynamics Setup** option from the **Subcase** node to define parameters related to dynamic loading. The **Dynamics Setup** dialog box will be displayed; refer to Figure-6.

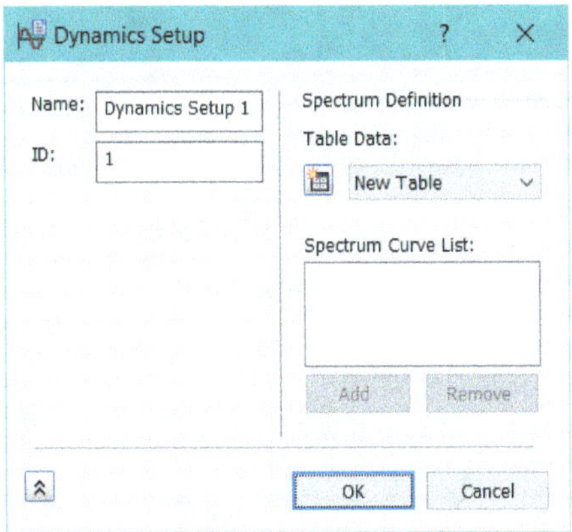

Figure-6. Dynamics Setup dialog box

- Click on **Define New Table** button from the **Table Data** area. The **Table Data** dialog box will be displayed; refer to Figure-7.

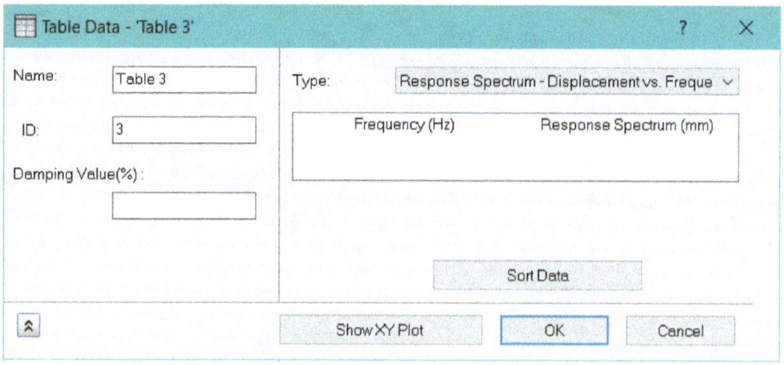

Figure-7. Table data for response spectrum

- Select the **Response Spectrum-Acceleration vs. Frequency** option from the **Type** drop-down and set the desired parameters for response spectrum; refer to Figure-8.

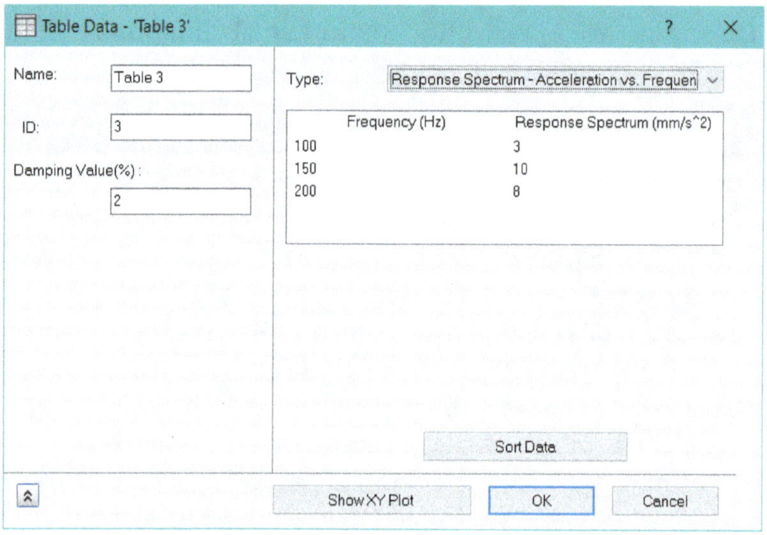

Figure-8. Parameters for response spectrum

- Click on the **OK** button from the dialog box.
- Select the newly created table data from the **Table Data** drop-down and click on the **Add** button. The selected spectrum curve table will be added in the **Spectrum Curve List** box; refer to Figure-9.

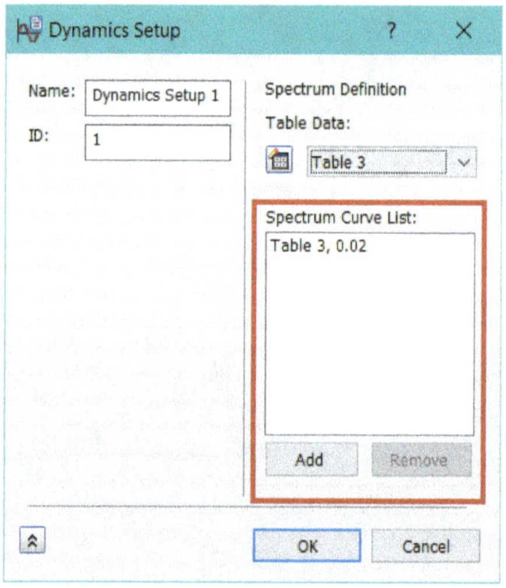

Figure-9. Spectrum curve added

- After setting the desired values, click on the **OK** button from the **Dynamics Setup** dialog box.
- Click on the **Constraints** tool from the **Setup** panel. The **Constraint** dialog box will be displayed with Response Spectrum constraint types.
- Select the desired check boxes from the dialog box to define how model will be constrained during the analysis.
- Set the other parameters as discussed earlier and click on the **Run** tool from the **Solve** panel of **Ribbon**. The analysis result will be displayed; refer to Figure-10.

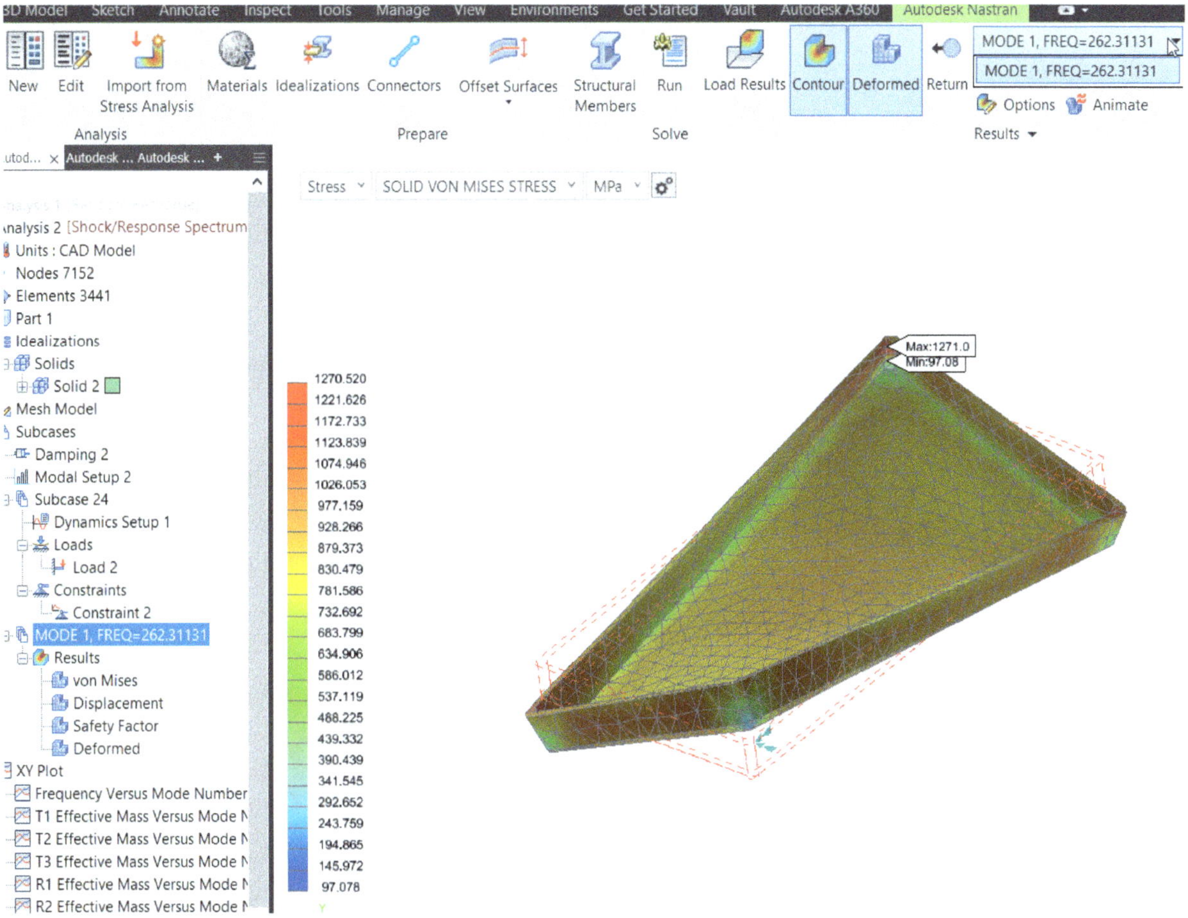

Figure-10. Result of shock analysis

For Student Notes

Chapter 10

Fatigue Analyses

Topics Covered

The major topics covered in this chapter are:

- ***Introduction.***
- ***Performing Multi-axial Fatigue Analysis***
- ***Defining Fatigue Setup Parameters***
- ***Performing Vibration Fatigue Analysis***

INTRODUCTION

Fatigue analyses are performed to check the effect of continuous loading and unloading of component for specified duration. There are two types of fatigue analyses available in Autodesk Inventor Nastran viz. Multi-Axial Fatigue and Vibration Fatigue. The procedures to perform these analyses are discussed next.

PERFORMING MULTI-AXIAL FATIGUE ANALYSIS

The Multi-axial fatigue analysis is performed when a structural load is applied and removed from the model continuously for long time. The procedure to perform multi-axial fatigue analysis is given next.

- Open the model on which you want to perform multi-axial fatigue analysis and start **Autodesk Inventor Nastran**. Linear static analysis will be activated automatically.
- Click on the **Edit** button from the **Analysis** panel of **Autodesk Inventor Nastran** tab in the **Ribbon**. The **Analysis** dialog box will be displayed.
- Select the **Multi-Axial Fatigue** option from the **Type** drop-down and set the desired parameters.
- Click on the **OK** button from the dialog box. The parameters for fatigue setup will be displayed; refer to Figure-1.

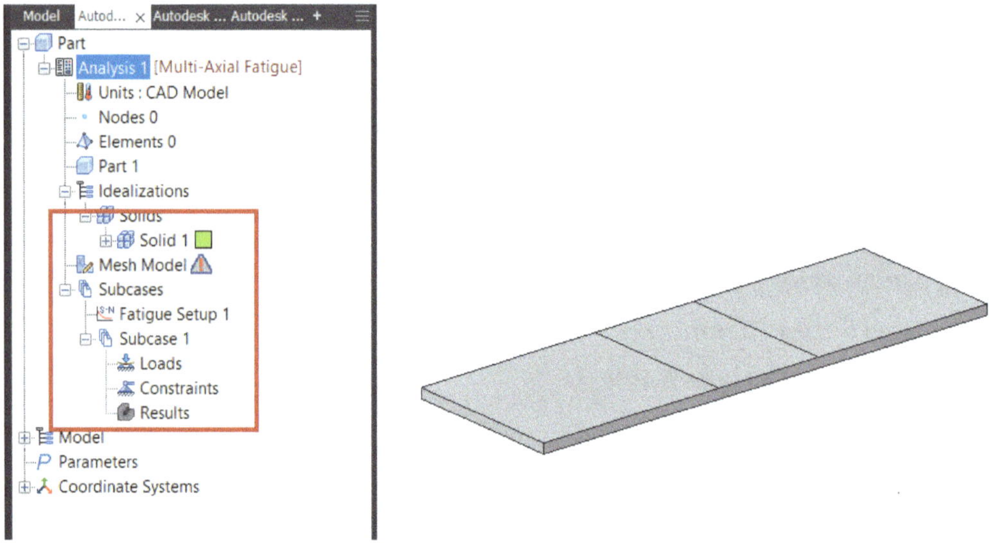

Figure-1. Parameters for multi-axial fatigue

Fatigue Parameters

- Double-click on **Fatigue Setup 1** option from **Subcases** node in **Model Tree**. The **Fatigue Setup** dialog box will be displayed; refer to Figure-2.

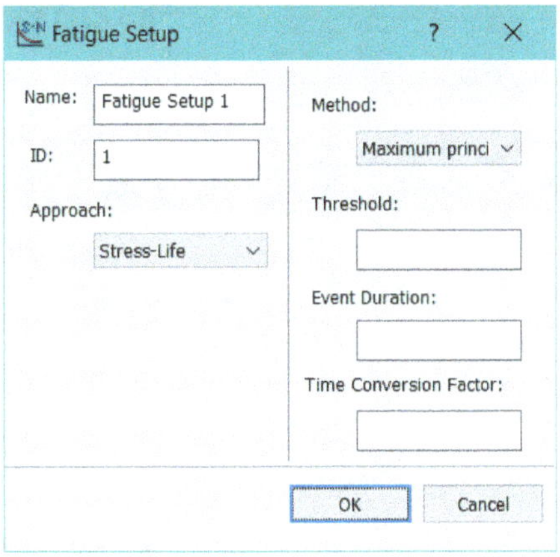

Figure-2. Fatigue Setup dialog box

- Select the desired option from the **Approach** drop-down. Select the **Stress-Life** option if you want to use SN data of material for calculating fatigue deformation. Select the **Strain-Life** option from the drop-down if you want to use EN data of material for calculating fatigue deformation.
- Select the desired option from the **Method** drop-down to define how life of part will be calculated.
- Specify the desired value of loading threshold (percentage) in **Threshold** edit box.
- Specify the desired value of fatigue study duration (in number of cycles) in the **Event Duration** edit box. Each cycle consists of one time load and unloading of acting loads
- Specify the desired value in **Time Conversion Factor** edit box to define the conversion of default time unit which is second to desired unit.
- Set the desired parameters as required and click on the **OK** button from the dialog box.
- Set the other parameters like idealization, load, constraint and click on the **Run** button from the **Solve** panel. The result of analysis will be displayed; refer to Figure-3.

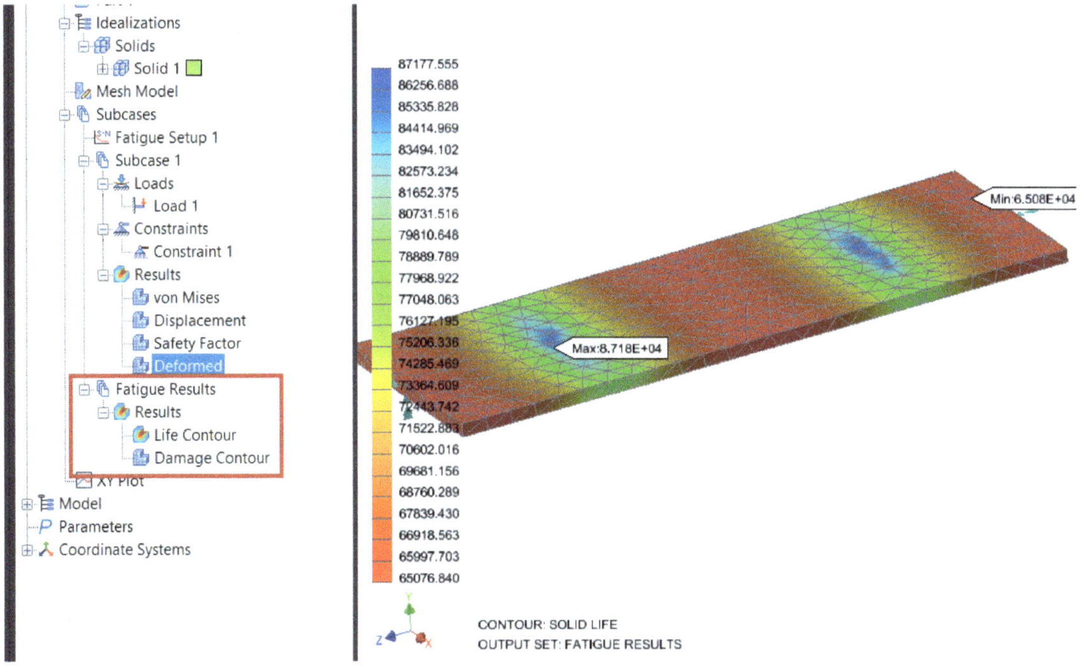

Figure-3. Fatigue analysis result

- Double-click on desired option in **Results** node. The respective result will be displayed.

PERFORMING VIBRATION FATIGUE ANALYSIS

The vibration fatigue analysis is performed when there is continuous loading and unloading of vibrational excitation. The procedure to perform this analysis is given next.

- Open the part on which you want to vibrational fatigue analysis and start Autodesk Inventor Nastran.
- Click on the **Edit** button from the **Analysis** panel of **Autodesk Inventor Nastran** tab in the **Ribbon**. The **Analysis** dialog box will be displayed.
- Select the **Vibration Fatigue** option from the **Type** drop-down. The options related to vibrational fatigue analysis will be displayed.
- Set the other parameters as discussed earlier and click on the **OK** button from the dialog box. The options for vibrational fatigue analysis will be displayed as shown in Figure-4.

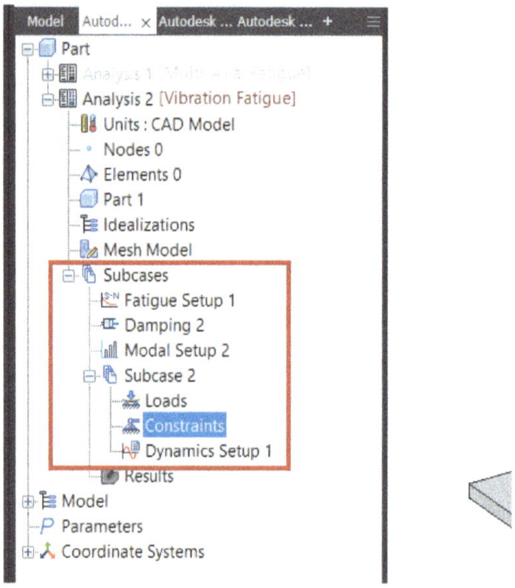

Figure-4. Options for vibration fatigue analysis

- Double-click on the parameters under **Subcases** node and set the desired values.
- Click on the **Run** button after setting all the parameters. The results will be displayed; refer to Figure-5.

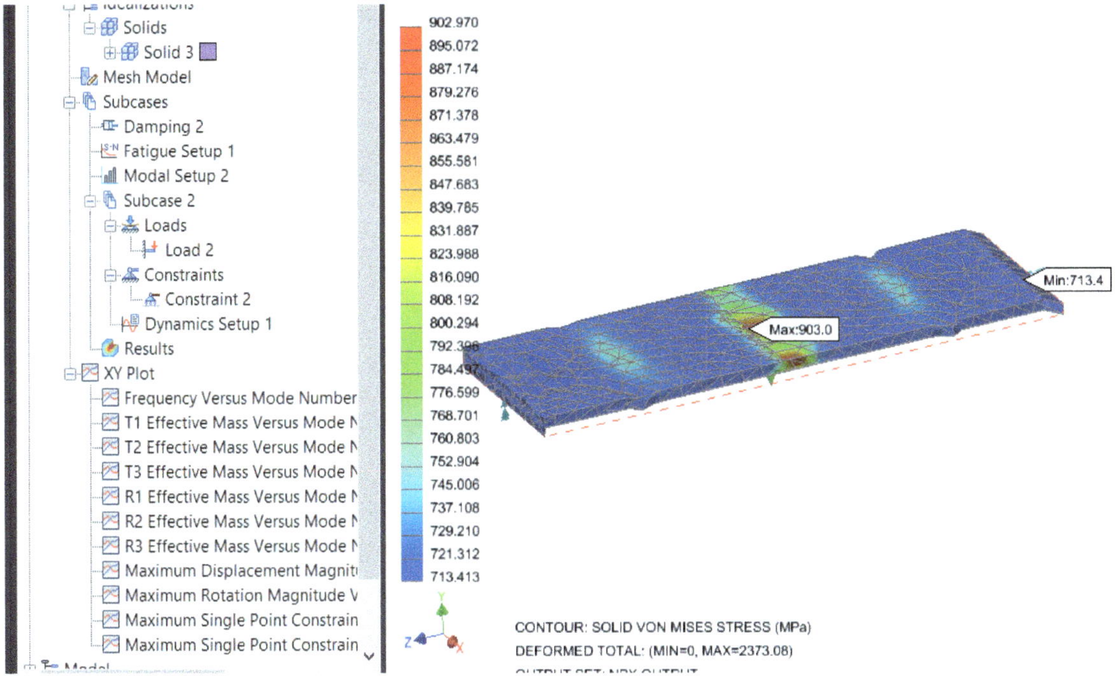

Figure-5. Result of vibration fatigue analysis

FOR STUDENT NOTES

Chapter 11

Heat Transfer Analyses

Topics Covered

The major topics covered in this chapter are:

- ***Introduction.***
- ***Performing Linear Steady State Heat Transfer Analysis***
- ***Performing Nonlinear Steady State Heat Transfer Analysis***
- ***Performing Nonlinear Transient Heat Transfer Analysis***

INTRODUCTION

The heat transfer analyses are performed to check the distribution of temperature over selected model which in turn can be used to check expansion of part due to heat. There are three analyses available in Autodesk Inventor Nastran to perform thermal analyses; Linear Steady State Heat Transfer, Nonlinear Steady State Heat Transfer, and Nonlinear Transient Heat Transfer. These analyses are discussed next.

PERFORMING LINEAR STEADY STATE HEAT TRANSFER ANALYSIS

The linear steady state heat transfer analysis is performed when thermal conductivity of material do not depend on temperature and radiation is not applied. The procedure to perform linear steady heat transfer analysis is given next.

- Open the model on which you want to perform linear steady state heat transfer analysis and start Autodesk Inventor Nastran environment.
- Click on the **Edit** button from the **Analysis** panel of **Autodesk Inventor Nastran** tab in the **Ribbon**. The **Analysis** dialog box will be displayed.
- Select the **Linear Steady State Heat Transfer** option from the **Type** drop-down. The options will be displayed as shown in Figure-1.

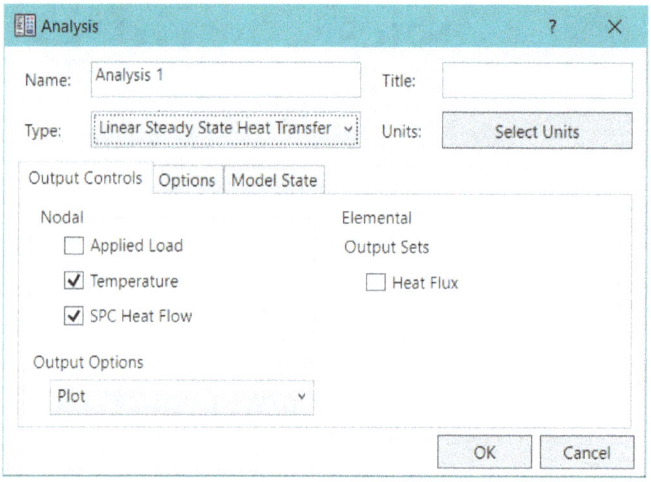

Figure-1. Linear Steady State Heat Transfer analysis options

- Select the **Applied Load** check box if you want to add applied load as output on nodes.
- If you want to generate heat flux as output on elements then select the **Heat Flux** check box from **Elemental** area of the dialog box.
- Click on the **Options** tab and select the desired options from the **Contact Type** drop-down. Note that bonded option is used to perfectly transfer heat between assembled components.
- After setting the desired parameters, click on the **OK** button from the dialog box. The options for linear steady state heat transfer analysis will be displayed; refer to Figure-2.

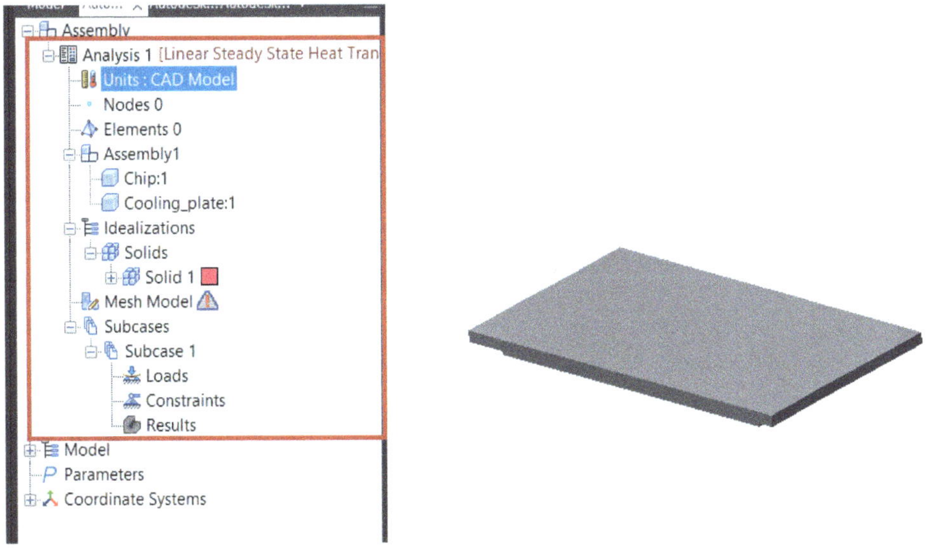

Figure-2. Options for linear steady state thermal analysis

- Click on the **Constraints** tool from the Setup panel of Autodesk Inventor Nastran tab in the Ribbon. The **Constraint** dialog box will be displayed; refer to Figure-3.

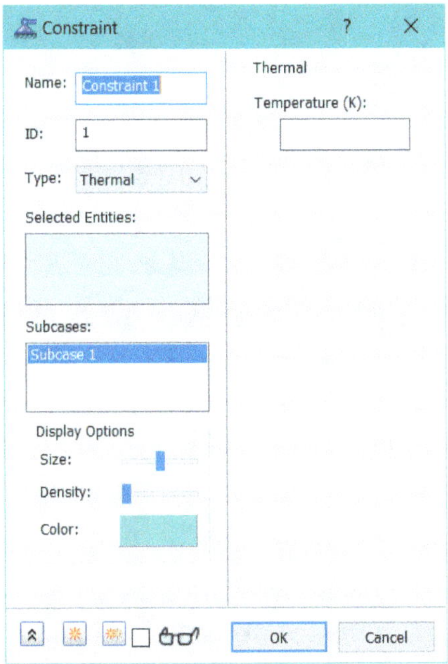

Figure-3. Constraint dialog box with thermal constraint

- Select the face/edge/vertex on which you want to specify fixed temperature.
- Specify the desired value of temperature in the **Temperature (K)** edit box and click on the **OK** button.
- Click on the **Loads** tool from the **Setup** panel of **Autodesk Inventor Nastran** tab in the **Ribbon**. The **Load** dialog box will be displayed.
- Select the **Heat Generation** option from the **Type** drop-down if you want to apply heat generation at selected body, face, edge, or vertex.
- Specify the desired value of heat generation in the **Magnitude** edit box and click on the **OK** button. Similarly, set the other thermal loads like heat convection, heat flux, radiation, and other parameters.

- Set the other parameters as required and click on the **Run** tool from the **Solve** panel of **Autodesk Inventor Nastran** tab in the **Ribbon**. The results will be displayed; refer to Figure-4.

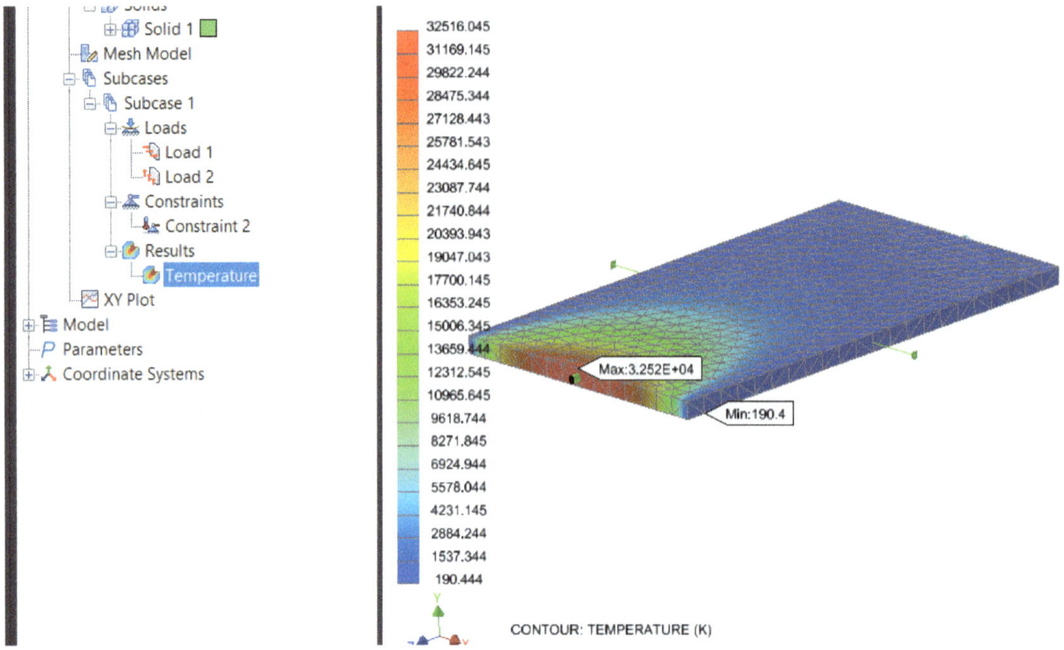

Figure-4. Thermal analysis result

Click on the **Options** tool from the **Results** panel and set the desired parameters to check results.

PERFORMING NONLINEAR STEADY STATE HEAT TRANSFER ANALYSIS

The Nonlinear steady state heat transfer analysis is performed when either thermal conductivity of part changes with temperature or radiation is applied on the model. The procedure to perform nonlinear steady state heat transfer analysis is given next.

- Open the part on which you want to perform analysis and start Autodesk Inventor Nastran environment.
- Click on the **Edit** tool from the **Analysis** panel of **Autodesk Inventor Nastran** tab in the **Ribbon**. The **Analysis** dialog box will be displayed.
- Select the **Nonlinear Steady State Heat Transfer** option from the **Type** drop-down and set the other parameters as required. Click on the **OK** button from the dialog box. The options related to nonlinear steady state thermal analysis will be displayed.
- Click on the **Loads** tool from the **Setup** panel of **Autodesk Inventor Nastran** tab in the **Ribbon**. The **Loads** dialog box will be displayed.
- Select **Initial Condition** option from the **Type** drop-down and **Temperature** option from the **Sub Type** drop-down.
- Specify the desired temperature value in the **Temperature** edit box and click on the **OK** button.
- Set the desired thermal constraint and loads as discussed earlier.

- After setting the desired parameters, click on the **Run** button from the **Solve** panel of **Autodesk Inventor Nastran** tab in the **Ribbon**. The analysis results will be displayed; refer to Figure-5.

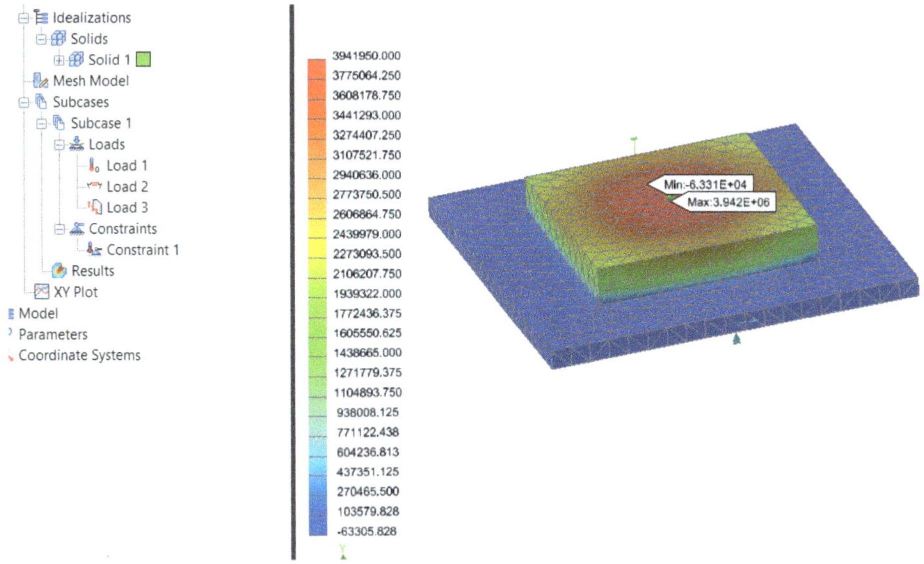

Figure-5. Nonlinear steady state thermal analysis result

PERFORMING NONLINEAR TRANSIENT HEAT TRANSFER ANALYSIS

Nonlinear transient heat transfer analysis is performed to check heat distribution at different time steps. The procedure to perform this analysis is given next.

- Open the part on which you want to perform the analysis and start **Autodesk Inventor Nastran** environment.
- Click on the **Edit** tool from the **Analysis** panel of **Autodesk Inventor Nastran** tab in the **Ribbon**. The **Analysis** dialog box will be displayed.
- Select the **Nonlinear Transient Heat Transfer** option from the **Type** drop-down and set the other options as required.
- Click on the **OK** button from the dialog box.
- Double-click on the **Time Step 1** option from the **Subcases** node. The **Time Step** dialog box will be displayed; refer to Figure-6.

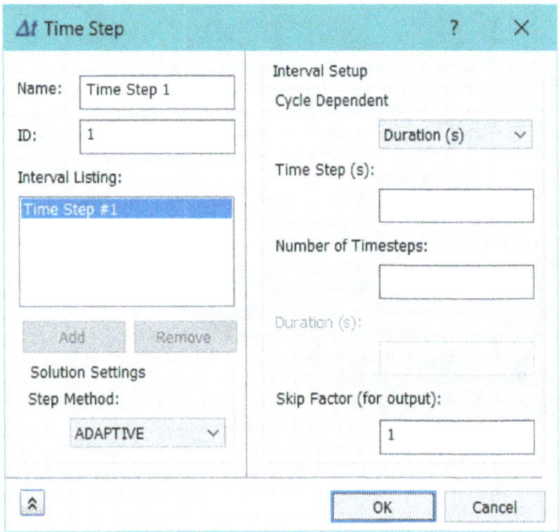

Figure-6. Time Step dialog box

- Specify the desired time duration and time steps as required in the dialog box, and click on the **OK** button from the dialog box.
- Set the other parameters like idealization, load, and constraint.
- Click on the **Run** tool from the **Solve** panel. The analysis result will be displayed; refer to Figure-7.

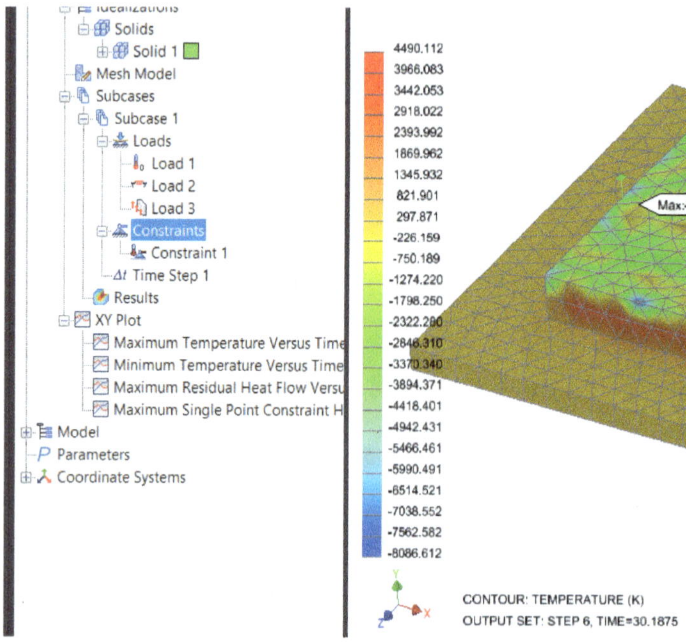

Figure-7. Result of nonlinear transient heat transfer analysis

Select the desired time step from the **Results** drop-down in **Results** panel of **Ribbon**.

PRACTICAL

Perform transient thermal analysis on the model with specified conditions as shown in Figure-8.

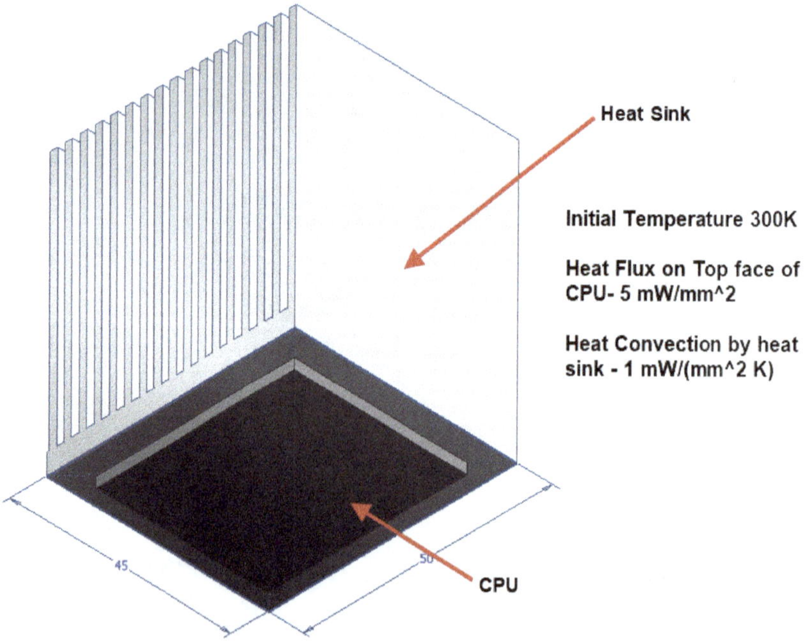

Figure-8. Model for Practice

For Student Notes

For Student Notes

Chapter 12

Explicit Analyses

Topics Covered

The major topics covered in this chapter are:

- ***Introduction.***
- ***Performing Explicit Dynamics Analysis***
- ***Performing Explicit Quasi-Static Analysis***

INTRODUCTION

The Explicit dynamics analysis is performed when severe load is applied on the model for very short time (in milli- or micro-seconds). Some of the examples where you should use explicit dynamic analyses are crash testing, metal forming, drop testing, explosion, on sports equipment, and so on.

The Explicit quasi-static analysis is used to analyze linear or nonlinear problems with time-dependent material changes like swelling in material, creep, viscoelasticity, and so on. The load is applied very slowing on the structure and the structure also deforms slowly. Note that effect of inertia is neglected for this analysis. The procedures to apply these analyses are discussed next.

PERFORMING EXPLICIT DYNAMICS ANALYSIS

The procedure to perform explicit dynamics analysis is given next.

- Open the model on which you want to perform explicit dynamics analysis and start Autodesk Inventor Nastran environment.
- Click on the **Edit** button from the **Analysis** panel in the **Autodesk Nastran Inventor** tab of **Ribbon**. The **Analysis** dialog box will be displayed.
- Select the **Explicit Dynamics** option from the **Type** drop-down and set other parameters as desired.
- Click on the **OK** button from the dialog box.
- Click on the **Materials** tool from the **Prepare** panel in the **Autodesk Inventor Nastran** tab of **Ribbon** and apply Steel AISI 1015 material.

Applying Shell Idealization

- Right-click on **Solid 1** from **Solids** node in **Idealization** category and click on the **Edit** option from the shortcut menu. The **Idealizations** dialog box will be displayed. Note that your component is a flat object with same thickness then you can convert solid object idealization to shell.
- Select the **Shell Elements** option from the **Type** drop-down and specify desired thickness for shell element; refer to Figure-1.

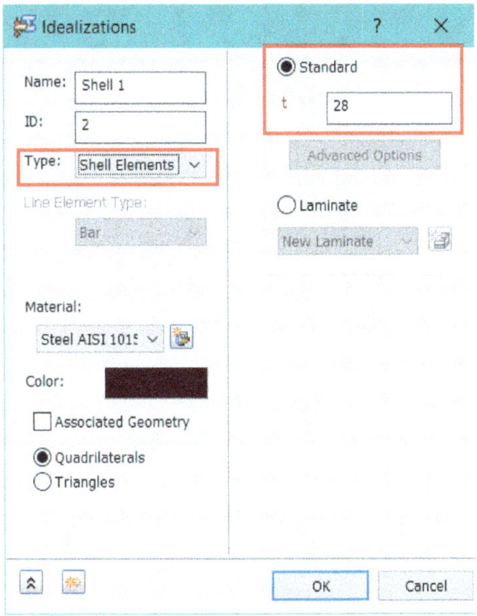

Figure-1. Specifying shell idealization

- Select the desired element type whether quadrilaterals or triangles, and click on the **OK** button.
- Click on the **Constraints** tool from the **Setup** panel in the **Autodesk Inventor Nastran** tab of **Ribbon** and apply desired constraints.

Applying Impulse Load

- Click on the **Loads** tool from the **Setup** panel of **Autodesk Inventor Nastran** tab in the **Ribbon**. The **Load** dialog box will be displayed.
- Set the parameters as desired in the dialog box and click on the **Define New Table** button from **Transient Table Data** area of the dialog box. The **Table Data** dialog box will be displayed.
- For an impulsive load, the load should be applied for a small instance of time and then it should be zero. So, specify the parameters accordingly in the table; refer to Figure-2.

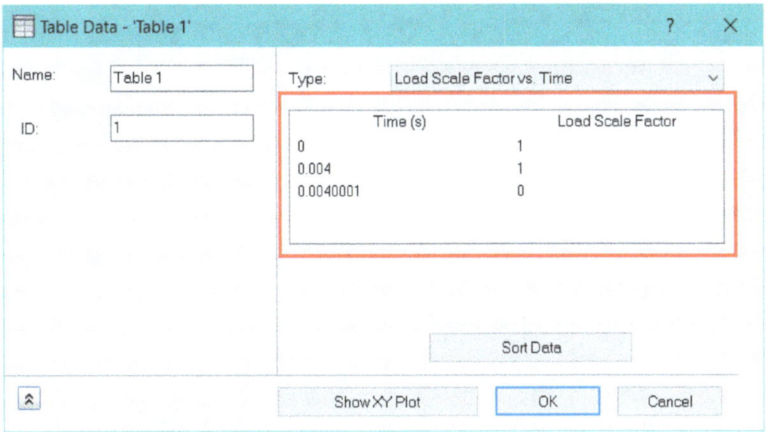

Figure-2. Table data specified

- Click on the **Show XY Plot** button from the dialog box to make sure the load is impulsive; refer to Figure-3.

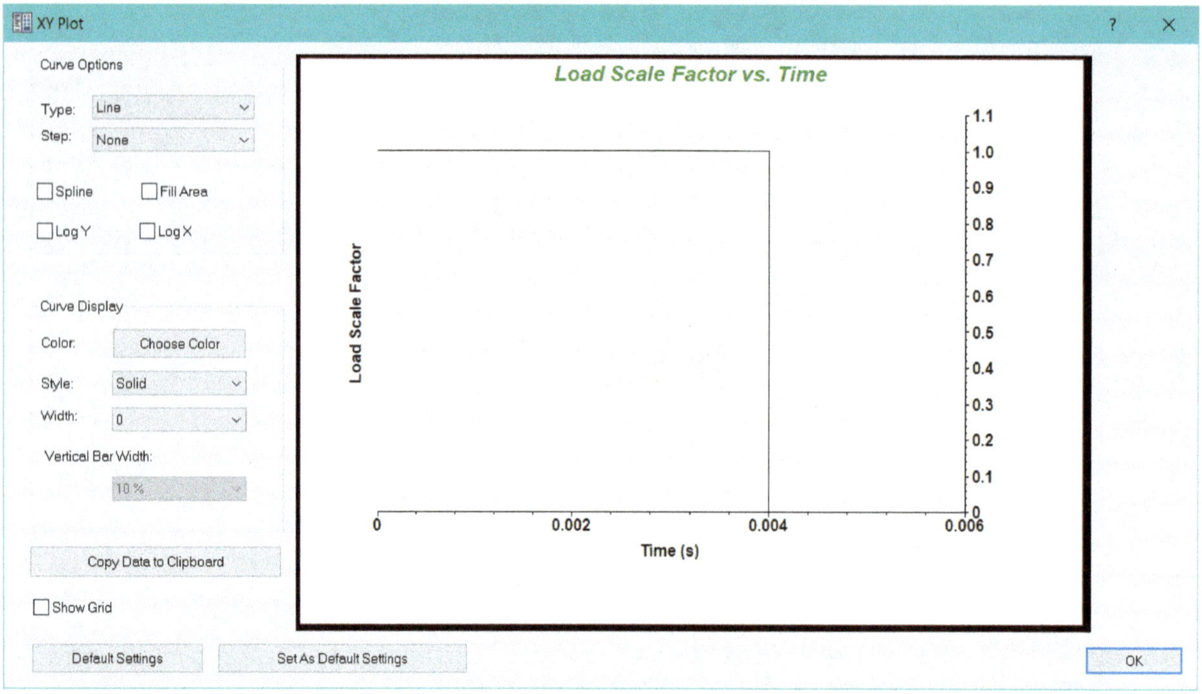

Figure-3. XY plot for impulsive load

- Click on the **OK** button from the **XY Plot** and **Table Data** dialog boxes to apply the parameters.
- Specify the other parameters as desired in the Load dialog box; refer to Figure-4 and click on the **OK** button.

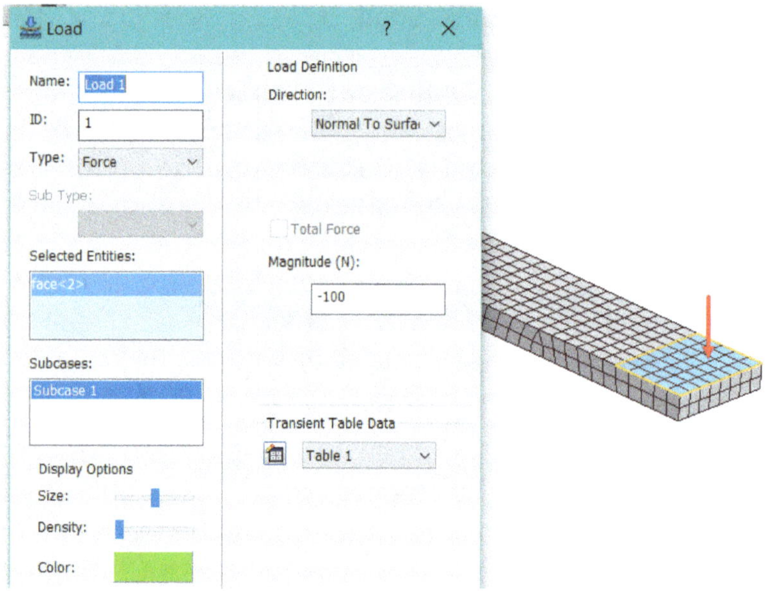

Figure-4. Applying impulsive load

- Click on the **Mesh Settings** button from the **Mesh** panel in the **Autodesk Inventor Nastran** tab in the **Ribbon**. The **Mesh Settings** dialog box will be displayed.
- Select the **Linear** option from the **Element Order** drop-down because we are using shell elements and **Parabolic** option is available only for solids. Set the other parameters as desired and click on the **Generate Mesh** button.
- Click on the **OK** button from the dialog box to exit.

- Double-click on the **Dynamic Setup 1** option from the **Subcase** node. The **Dynamic Setup** dialog box will be displayed.
- Set the time duration, interval and other parameters for dynamic study; refer to Figure-5.

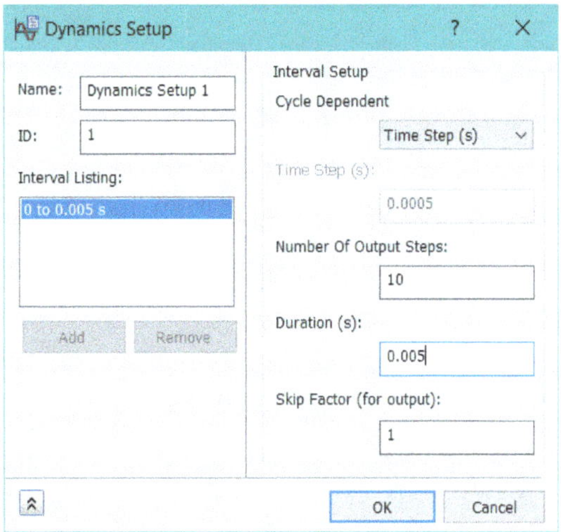

Figure-5. Dynamics Setup dialog box4

- Click on the **OK** button from the dialog box to apply settings.
- Click on the **Run** tool from the **Solve** panel in the **Autodesk Inventor Nastran** tab of **Ribbon**. The result will be displayed.
- Select the desired time point at which you want to check the results from the **Results** drop-down in the **Results** panel of **Ribbon**; refer to Figure-6.

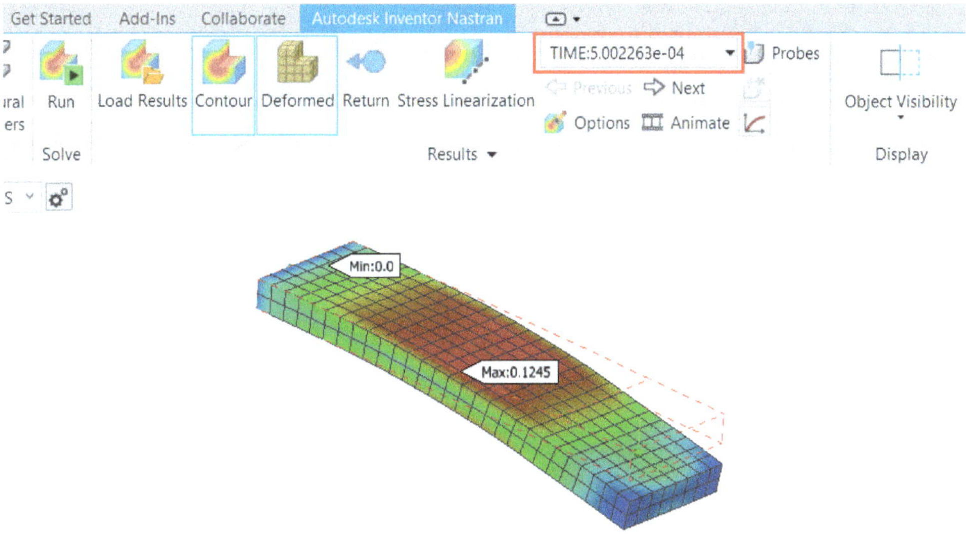

Figure-6. Tiime point selected for result

PERFORMING EXPLICIT QUASI-STATIC ANALYSIS

The procedure to perform a quasi-static analysis is given next.

- Click on the **New** tool from the **Analysis** panel in the **Autodesk Inventor Nastran** tab of **Ribbon**. The **Analysis** dialog box will be displayed.
- Select the **Explicit Quasi-Static** option from the **Type** drop-down and set the other parameters as desired; refer to Figure-7.

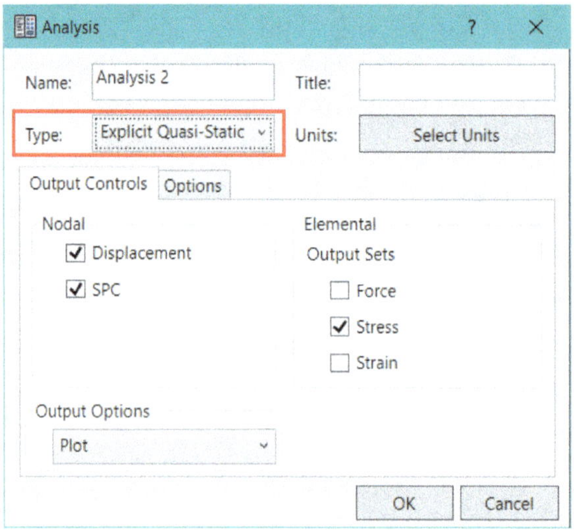

Figure-7. Options for explicit quasi-static analysis

- Click on the **OK** button from the dialog box. The parameters of analysis will be displayed in the **Model Browser**; refer to Figure-8.

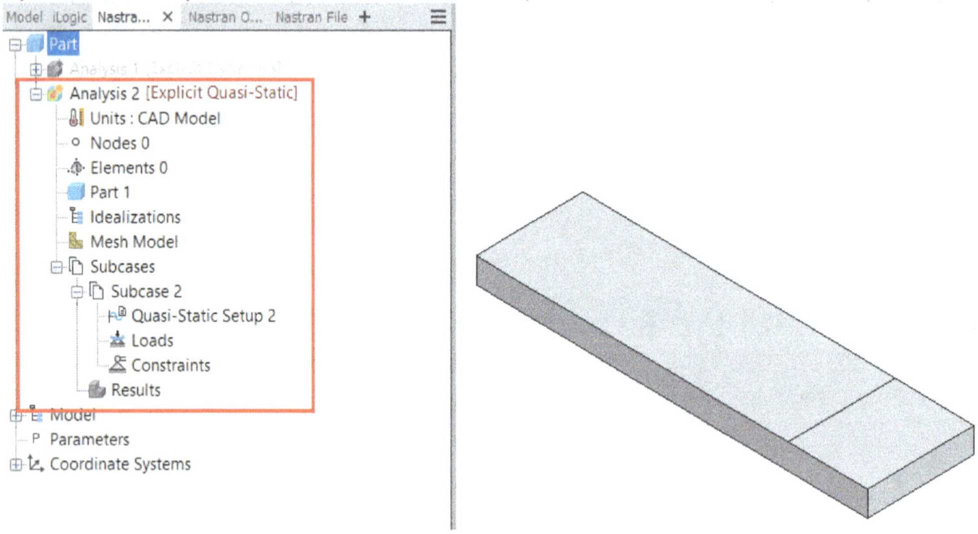

Figure-8. Parameters for explicit quasi-static analysis

- Double-click on **Quasi-Static Setup** option from the **Subcase** node in the **Model Browser**. The **Quasi-Static Setup** dialog box will be displayed; refer to Figure-9.

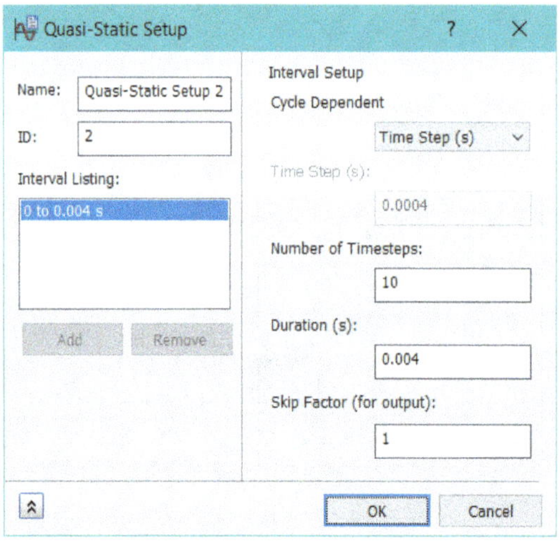

Figure-9. Quasi-Static Setup dialog box

- Set the desired time period with time steps in the dialog box and click on the **OK** button.
- Set the other parameters as desired and click on the **Run** button to perform the analysis.

FOR STUDENT NOTES

For Student Notes

Chapter 13

Fundamentals of FEA

Topics Covered

The major topics covered in this chapter are:

- ***Introduction to FEA.***
- ***General Description of FEA***
- ***Solving a problem with FEA***
- ***FEA V/s Classical Methods***
- ***FEA V/s Finite Different Method***
- ***Need for Studying FEM***
- ***Refining Mesh***
- ***Higher order element V/s Refined Mesh***

INTRODUCTION

The finite element analysis is a numerical technique. In this method all the complexities of the problems, like varying shape, boundary conditions and loads are maintained as they are but the solutions obtained are approximate. Because of its diversity and flexibility as an analysis tool, it is receiving much attention in engineering. The fast improvements in computer hardware technology and slashing of cost of computers have boosted this method, since the computer is the basic need for the application of this method. A number of popular brand of finite element analysis packages are now available commercially. Some of the popular packages are SolidWorks Simulation, NASTRAN, NISA, and ANSYS. Using these packages, one can analyze several complex structures.

The finite element analysis originated as a method of stress analysis in the design of aircrafts. It started as an extension of matrix method of structural analysis. Today this method is used not only for the analysis in solid mechanics, but even in the analysis of fluid flow, heat transfer,electric and magnetic fields and many others. Civil engineers use this method extensively for the analysis of beams, space frames, plates, shells, folded plates, foundations, rock mechanics problems and seepage analysis of fluid through porous media. Both static and dynamic problems can be handled by finite element analysis. This method is used extensively for the analysis and design of ships, aircraft, space crafts, electric motors, and heat engines.

GENERAL DESCRIPTION OF THE METHOD

In engineering problems there are some basic unknowns. If they are found, the behavior of the entire structure can be predicted. The basic unknowns or the field variables which are encountered in the engineering problems are displacements in solid mechanics, velocities in fluid mechanics, electric and magnetic potentials in electrical engineering and temperatures in heat flow problems.

In a continuum, these unknowns are infinite. The finite element procedure reduces such unknowns to a finite number by dividing the solution region into small parts called elements and by expressing the unknown field variables in terms of assumed approximating functions (Interpolating functions/Shape functions) within each element. The approximating functions are defined in terms of field variables of specified points called nodes or nodal points. Thus in the finite element analysis, the unknowns are the field variables of the nodal points. Once these are found the field variables at any point can be found by using interpolation functions.

After selecting elements and nodal unknowns next step in finite element analysis is to assemble element properties for each element. For example, in solid mechanics, we have to find the force-displacement i.e. stiffness characteristics of each individual element. Mathematically this relationship is of the form

$$[k]_e \{\delta\}_e = \{F\}_e$$

where $[k]_e$ is element stiffness matrix, $\{\delta\}_e$ is nodal displacement vector of the element and $\{F\}_e$ is nodal force vector. The element of stiffness matrix k_{ij} represent the force in coordinate direction 'i' due to a unit displacement in coordinate direction 'j'. Four methods are available for formulating these element properties viz. direct approach, variational approach, weighted residual approach and energy balance approach. Any one of these methods can be used for assembling element properties. In solid mechanics variational approach is commonly employed to assemble stiffness matrix and nodal force vector (consistent loads).

Element properties are used to assemble global properties/structure properties to get system equations $[k] \{\delta\} = \{F\}$. Then the boundary conditions are imposed. The solution of these simultaneous equations give the nodal unknowns. Using these nodal values additional calculations are made to get the required values e.g. stresses, strains, moments, etc. in solid mechanics problems.

Thus the various steps involved in the finite element analysis are:

(i) Select suitable field variables and the elements.
(ii) Discritise the continua.
(iii) Select interpolation functions.
(iv) Find the element properties.
(v) Assemble element properties to get global properties.
(vi) Impose the boundary conditions.
(vii) Solve the system equations to get the nodal unknowns.
(viii) Make the additional calculations to get the required values.

A BRIEF EXPLANATION OF FEA FOR A STRESS ANALYSIS PROBLEM

The steps involved in finite element analysis are clarified by taking the stress analysis of a tension strip with fillets (refer Figure-1). In this problem stress concentration is to be studies in the fillet zone. Since the problem is having symmetry about both x and y axes, only one quarter of the tension strip may be considered as shown in Figure-2. About the symmetric axes, transverse displacements of all nodes are to be made zero. The various steps involved in the finite element analysis of this problem are discussed below:

Step 1: Four noded isoparametric element is selected for the analysis (However note that 8 noded isoparametric element is ideal for this analysis). The four noded iso-parametric element can take quadrilateral shape also as required for elements 12, 15, 18, etc. As there is no bending of strip, only displacement continuity is to be ensured but not the slope continuity. Hence displacements of nodes in x and y directions are taken as basic unknowns in the problem.

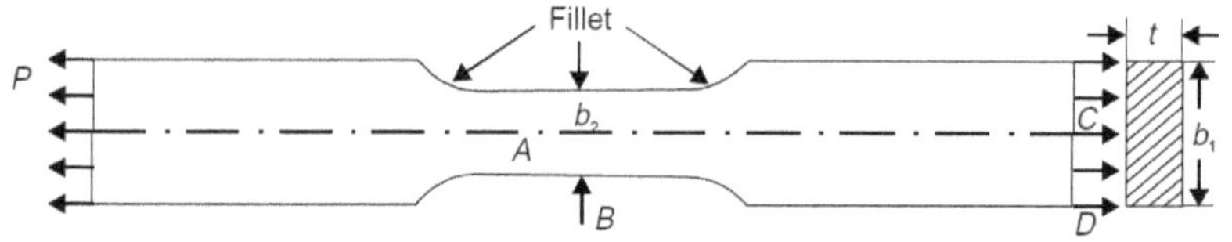

Figure-1. Tension Strip with Fillet

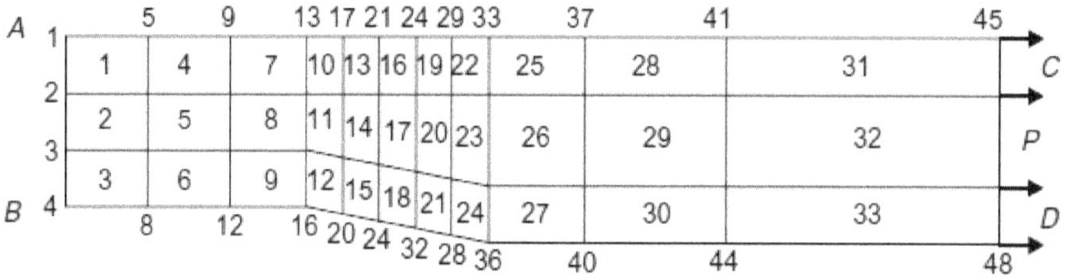

Figure-2. Descritization of quater of tension strip

Step 2: The portion to be analysed is to be discretised. Figure-2 shows discretised portion. For this 33 elements have been used. There are 48 nodes. At each node unknowns are x and y components of displacements. Hence in this problem total unknowns (displacements) to be determined are 48 × 2 = 96.

Step 3: The displacement of any point inside the element is approximated by suitable functions in terms of the nodal displacements of the element. For the typical element (Fig. 1.3 b), displacements at P are

$$u = \Sigma N_i u_i = N_1 u_1 + N_2 u_2 + N_3 u_3 + N_4 u_4$$

and

$$v = \Sigma N_i v_i = N_1 v_1 + N_2 v_2 + N_3 v_3 + N_4 v_4 \quad ...(1.2)$$

The approximating functions N_i are called shape functions or interpolation functions. Usually they are derived using polynomials.

Step 4: Now the stiffness characters and consistent loads are to be found for each element. There are four nodes and at each node degree of freedom is 2. Hence degree of freedom in each element is 4 × 2 = 8. The relationship between the nodal displacements and nodal forces is called element stiffness characteristics. It is of the form

$[k]_e \{\delta\}_e = \{F\}_e$, as explained earlier.

For the element under consideration, k_e is 8 × 8 matrix and δ_e and F_e are vectors of 8 values. In solid mechanics element stiffness matrix is assembled using variational approach i.e. by minimizing potential energy.

If the load is acting in the body of element or on the surface of element, its equivalent at nodal points are to be found using variational approach, so that right hand side of the above expression is assembled. This process is called finding consistent loads.

Step 5: The structure is having 48 × 2 = 96 displacement and load vector components to be determined.

Hence global stiffness equation is of the form

$$\underset{96 \times 96}{[k]} \quad \underset{96 \times 1}{\{\delta\}} = \underset{96 \times 1}{\{F\}}$$

Each element stiffness matrix is to be placed in the global stiffness matrix appropriately. This process is called assembling global stiffness matrix. In this problem force vector F is zero at all nodes except at nodes 45, 46, 47 and 48 in x direction. For the given loading nodal equivalent forces are found and the force vector F is assembled.

Step 6: In this problem, due to symmetry transverse displacements along AB and BC are zero. The system equation [k] {δ } = {F} is modified to see that the solution for {δ} comes out with the above values. This modification of system equation is called imposing the boundary conditions.

Step 7: The above 96 simultaneous equations are solved using the standard numerical procedures like Gauss elimination or Choleski's decomposition techniques to get the 96 nodal displacements.

Step 8: Now the interest of the analyst is to study the stresses at various points. In solid mechanics the relationship between the displacements and stresses are well established. The stresses at various points of interest may be found by using shape functions and the nodal displacements and then stresses calculated. The stress concentrations may be studies by comparing the values obtained at various points in the fillet zone with the values at uniform zone, far away from the fillet (which is equal to P/b_2t).

FINITE ELEMENT METHOD V/S CLASSICAL METHODS

1. In classical methods exact equations are formed and exact solutions are obtained where as in finite element analysis exact equations are formed but approximate solutions are obtained.

2. Solutions have been obtained for few standard cases by classical methods, where as solutions can be obtained for all problems by finite element analysis.

3. Whenever the following complexities are faced, classical method makes the drastic assumptions' and looks for the solutions:
(a) Shape
(b) Boundary conditions
(c) Loading

Figure-3 shows such cases in the analysis of slabs (plates).
To get the solution in the above cases, rectangular shapes, same boundary condition along a side and regular equivalent loads are to be assumed. In FEM no such assumptions are made. The problem is treated as it is.

4. When material property is not isotropic, solutions for the problems become very difficult in classical method. Only few simple cases have been tried successfully by researchers. FEM can handle structures with anisotropic properties also without any difficulty.

5. If structure consists of more than one material, it is difficult to use classical method, but finite element can be used without any difficulty.

6. Problems with material and geometric non-linearities can not be handled by classical methods.
There is no difficulty in FEM.

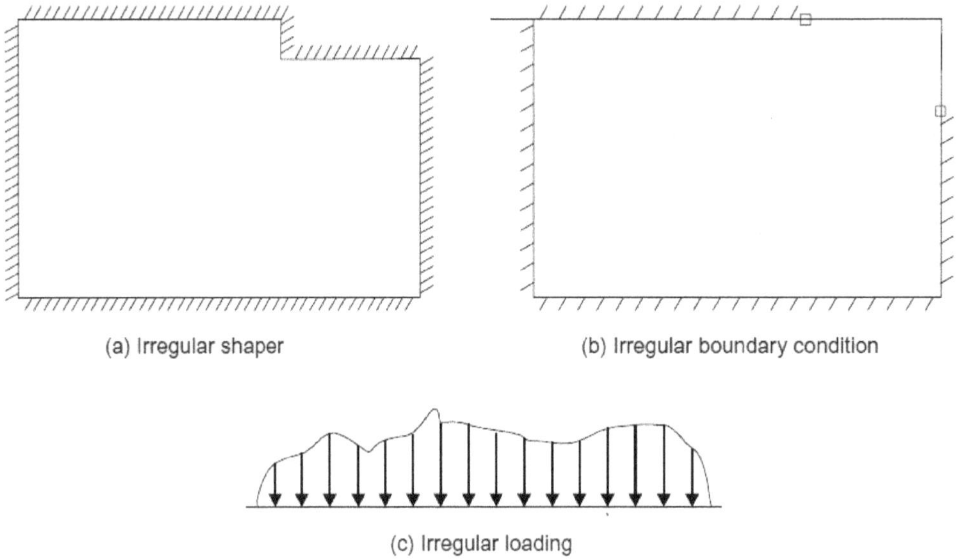

(a) Irregular shaper

(b) Irregular boundary condition

(c) Irregular loading

Figure-3. Analysis of slab

Hence FEM is superior to the classical methods only for the problems involving a number of complexities which cannot be handled by classical methods without making drastic assumptions. For all regular problems, the solutions by classical methods are the best solutions. Infact, to check the validity of the FEM programs developed, the FEM solutions are compared with the solutions by classical methods for standard problems.

FEM VS FINITE DIFFERENCE METHOD (FDM)

1. FDM makes pointwise approximation to the governing equations i.e. it ensures continuity only at the node points. Continuity along the sides of grid lines are not ensured. FEM make piecewise approximation i.e. it ensures the continuity at node points as well as along the sides of the element.

2. FDM do not give the values at any point except at node points. It do not give any approximating function to evaluate the basic values (deflections, in case of solid mechanics) using the nodal values. FEM can give the values at any point. However the values obtained at points other than nodes are by using suitable interpolation formulae.

3. FDM makes stair type approximation to sloping and curved boundaries as shown in Figure-4. FEM can consider the sloping boundaries exactly. If curved elements are used, even the curved boundaries can be handled exactly.

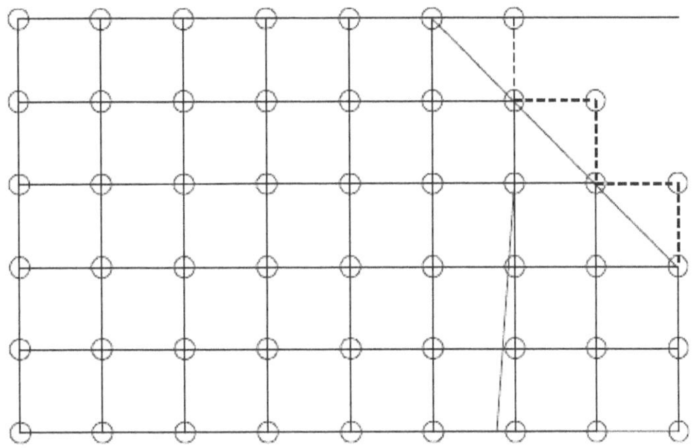

Figure-4. FDM approximation of shape

4. FDM needs larger number of nodes to get good results while FEM needs fewer nodes.

5. With FDM fairly complicated problems can be handled where as FEM can handle all complicated problems.

NEED FOR STUDYING FEM

Now, a number of users friendly packages are available in the market. Hence one may ask the question 'What is the need to study FEA?'.

The finite element knowledge makes a good engineer better while just user without the knowledge of FEA may produce more dangerous results. To use the FEA packages properly, the user must know the following points clearly:

1. Which elements are to be used for solving the problem in hand.
2. How to discritise to get good results.
3. How to introduce boundary conditions properly.

4. How the element properties are developed and what are their limitations.
5. How the displays are developed in pre and post processor to understand their limitations.
6. To understand the difficulties involved in the development of FEA programs and hence the need for checking the commercially available packages with the results of standard cases.

Unless user has the background of FEA, he may produce worst results and may go with overconfidence. Hence it is necessary that the users of FEA package should have sound knowledge of FEA.

Warning To Fea Package Users

When hand calculations are made, the designer always gets the feel of the structure and get rough idea about the expected results. This aspect cannot be ignored by any designer, whatever be the reliability of the program, a complex problem may be simplified with drastic assumptions and FEA results obtained. Check whether expected trend of the result is obtained. Then avoid drastic assumptions and get more refined results with FEA package. User must remember that structural behavior is not dictated by the computer programs. Hence the designer should develop feel of the structure and make use of the programs to get numerical results which are close to structural behavior.

One of the main concern for using FEA applications is selection of elements, in other words discretization. The process of modeling a structure using suitable number, shape, and size of the elements is called discretization. The modeling should be good enough to get the results as close to actual behavior of the structure as possible.

In a structure, we come across the following types of discontinuities:

(a) Geometric
(b) Load
(c) Boundary conditions
(d) Material.

Geometric Discontinuities

Wherever there is sudden change in shape and size of the structure there should be a node or line of nodes. Figure-5 shows some of such situations.

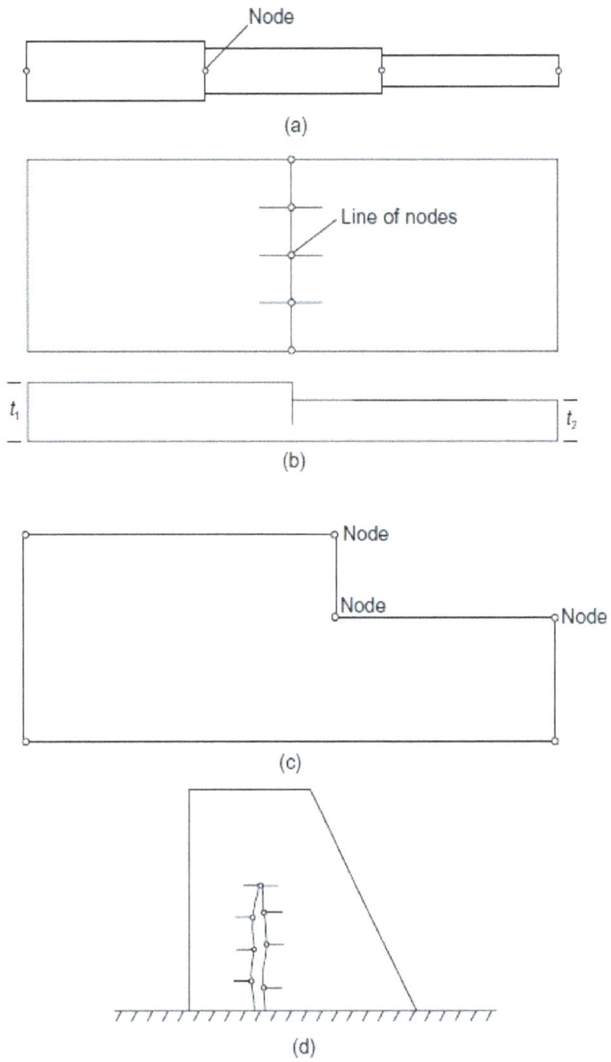

Figure-5. Geometric Discontinuity in the part

Discontinuity of Loads

Concentrated loads and sudden change in the intensity of uniformly distributed loads are the sources of discontinuity of loads. A node or a line of nodes should be there to model the structure. Some of these situations are shown in Figure-6.

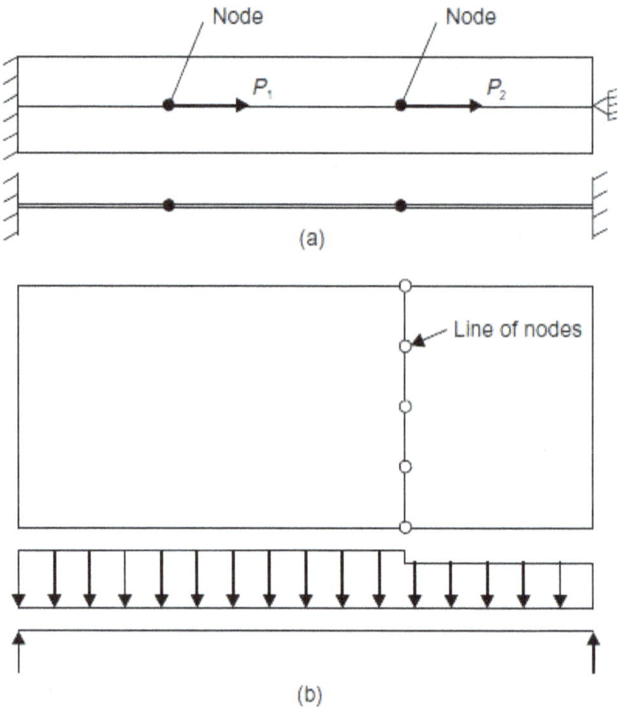

Figure-6. FEM Model and slab with different UDL

Discontinuity of Boundary conditions

If the boundary condition for a structure suddenly change we have to discretize such that there is node or a line of nodes. This type of situations are shown in Figure-7.

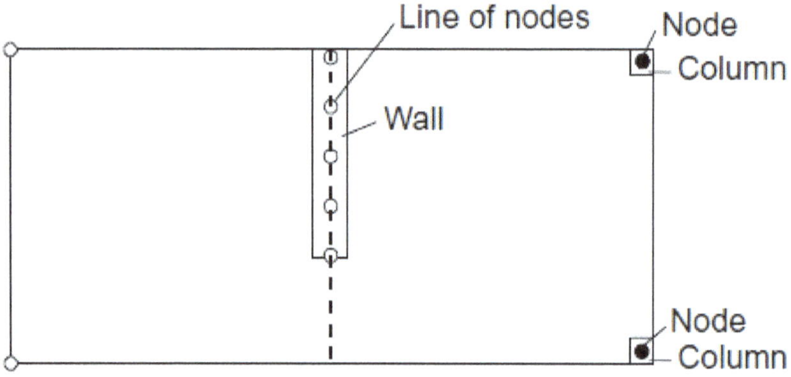

Figure-7. Slab with intermediate wall and columns

Material Discontinuity

Node or node lines should appear at the places where material discontinuity is seen; refer to Figure-8.

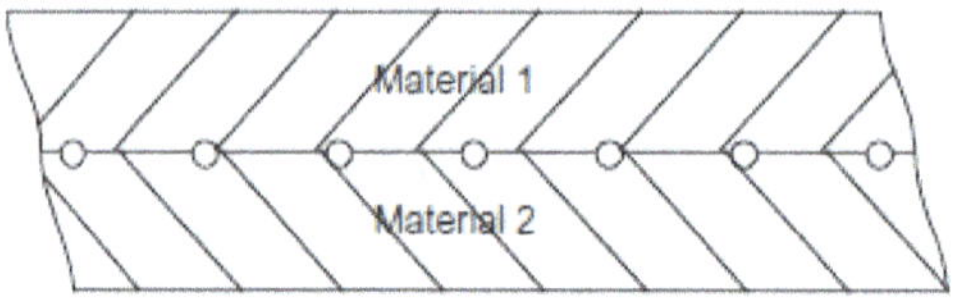

Figure-8. Material Discontinuity

REFINING MESH

To get better results the finite element mesh should be refined in the following situations:

(a) To approximate curved boundary of the structure
(b) At the places of high stress gradients.

Such a situation is shown in Figure-9.

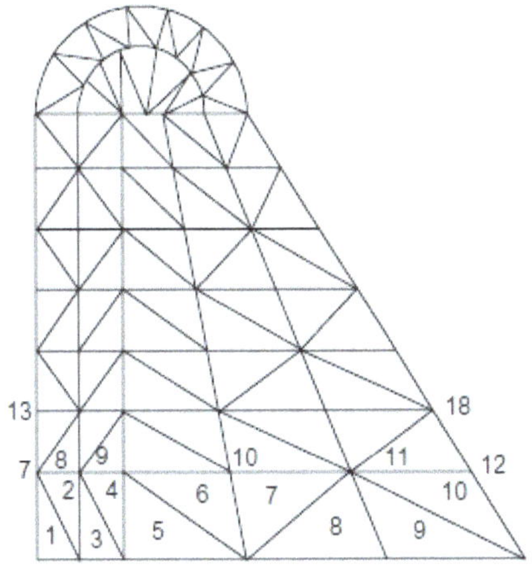

Figure-9. Refined mesh near curved boundary

Use Of Symmetry

Wherever there is symmetry in the problem, it should be made use. By doing so, lot of memory requirement is reduced or in other words we can use more elements (refined mesh) for the same capacity of computer memory. When symmetry is to be used, it is to be noted that at right angles to the line of symmetry displacement is zero.

HIGHER ORDER ELEMENTS V/S REFINED MESH

Accuracy of calculation increases if higher order elements are used. Accuracy can also be increased by using more number of elements. Limitation on use of number of elements comes from the total degrees of freedom the computer can handle. The limitation may be due to cost of computation time also. Hence to use higher order elements we have to use less number of such elements. The question arises whether to use less number of higher order elements or more number of lower order elements for the same total degree of freedom. There are some studies in this matter keeping degree of accuracy per unit cost as the selection criteria. However the cost of calculation is coming down so much that such studies are not relevant today. Accuracy alone should be selection criteria which may be carried out initially on the simplified problem and based on it element may be selection for detailed study.

For Student Notes

Practice Questions

Topics Covered

The major topics covered in this chapter are:

- ***Practice Questions***

PROBLEM 1

Sometimes we see people fighting and in some extreme cases, fighting with the help of Baseball bats. We are not here to discuss concepts of fighting but once the fighting is over, everybody is busy to see effect of baseball bat on humans. Due to unknown elastic property of human body, we can not perform analysis to find effect of baseball bat on humans, but here we can check the effect of force exerted on the baseball bat due to some freaks; refer to Figure-1.

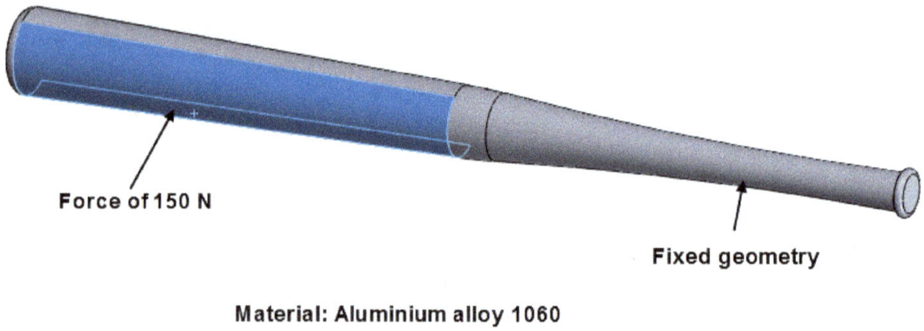

Figure-1. Problem 1

We assume it to be a static analysis. Check whether the bat will break or not.

PROBLEM 2

The wall of a house is 7 m wide, 6 m high and 0.3 m thick made up of brick. The thermal conductivity of brick is k = 0.6 W/m.K. If the inside temperature of wall is 16 °C and outside temperature is 6 °C then find out the heat lost from inside wall to outside wall in a steady state condition. The model is given in Figure-2.

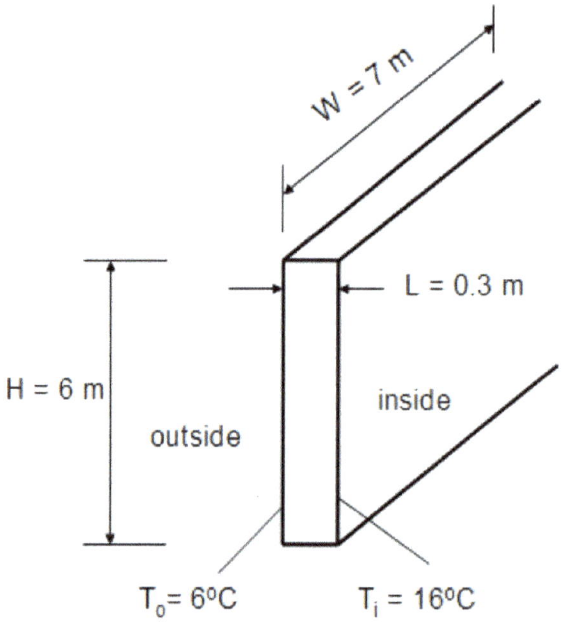

Figure-2. Problem 2

Note that to perform this thermal analysis, you need to create a new material with thermal properties as follow:

Thermal Conductivity = 0.6 W/(m.K)
Specific heat = 1214 J/(kg.K)
Mass density = 2405 kg/m^3

PROBLEM 3

An axial pull of **35kN** is applied at the ends of a shaft as shown in Figure-3. If the material is **Alloy Steel** then find the total elongation in the shaft.

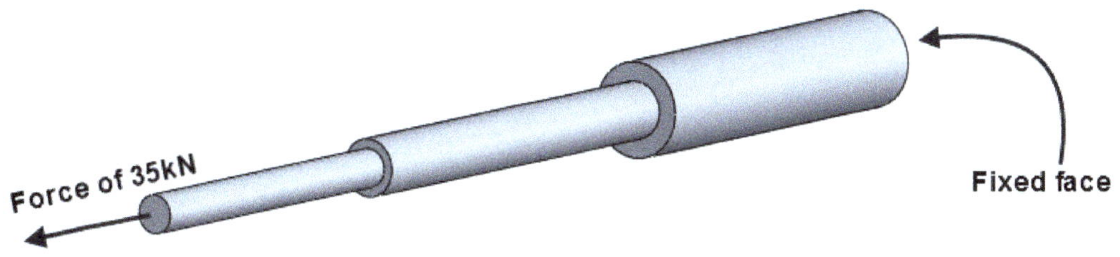

Figure-3. Problem 3 shaft

PROBLEM 4

A cast iron link is required to transmit a steady tensile load of 45kN. The initial boundary conditions are given in Figure-4.

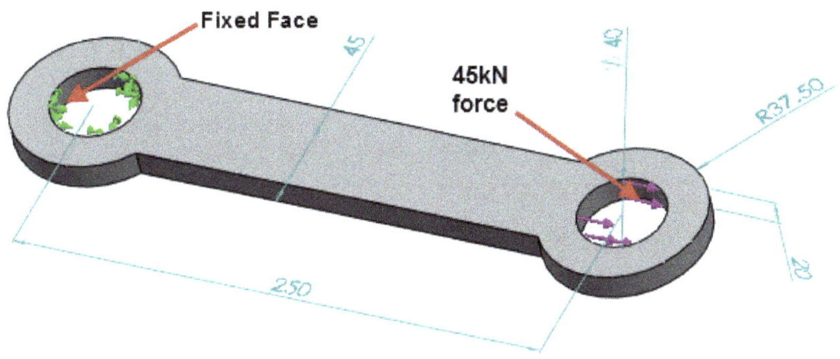

Figure-4. Problem of link

Find out the optimum thickness of link so that factor of safety is 3.

PROBLEM 5

A part of car jack assembly is displayed in Figure-5 with its boundary conditions. Optimize the assembly with respect to material cost. Note that all the contacts are no penetration as you find in a real car jack.

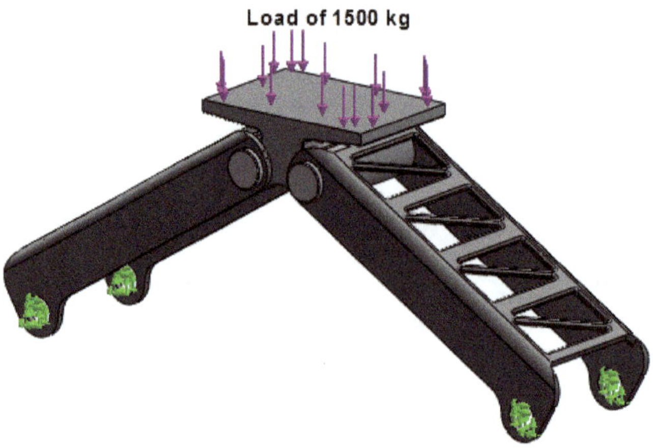

Figure-5. Problem 5

Problem 6

An assembly of cooling fins and CPU is displayed in Figure-6 with boundary conditions. The bulk temperature is 300K and operating temperature band for CPU is 304K to 354 K. Check if fins are enough to dissipate the heat. If not then perform modifications in fin to dissipate heat. Note that fins and heat sink are perfectly bonded for heat transfer.

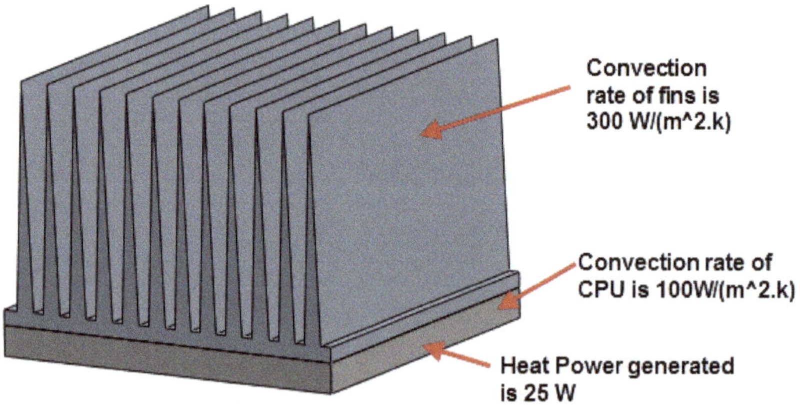

Figure-6. Problem 6

Index

Symbols

2D triangular shell elements 2-44
3D tetrahedral solid elements 2-44

A

Advanced Mesh Settings dialog box 2-45
Advanced Options button 2-25, 2-42
Advanced Settings button 7-5
Analysis dialog box 3-2
Animate button 2-53
Animation Options tab 2-52
Applying Contacts 3-8
Auto tool 2-39

B

Bearing Load option 2-30
Body Temperature option 2-33
Bolt option 2-17
Bonded option 2-41
buckling 5-2
Buckling analysis 5-3
Buckling Analysis 1-4
buckling mode shapes 5-2

C

Cable option 2-14
Choleski's decomposition 13-5
Cluster between Modes check box 8-6
Coefficient of Friction edit box 2-42
Conduction 1-8
Connector dialog box 2-13
Connectors tool 2-13
consistent loads 13-5
Constraint dialog box 2-23
Constraints tool 1-11, 2-23
Contact Parameters dialog box 2-42
Contact Type drop-down 2-8, 2-41, 11-2
Contour Options check box 2-50
Convection 1-8
Convection Coefficient vs. Temperature option 2-35
CQC option 9-5
Create AVI button 2-53
Creating New Material 2-10
creep analysis 3-12
critical buckling factors 5-2
cycle 1-3

D

Damping , 4-2
Damping dialog box 6-3
Default Settings tool 2-2
Define New Table button 2-34
Deform Options check box 2-50
Degrees of Freedom area 2-23
Dependent Term drop-down 7-5
Direct Frequency Response analysis 8-2
Direct Frequency Response Analysis 1-7
Direct Frequency Response option 8-2
Direct Transient Response Analysis 1-6, 6-2
Disable option 2-41
Discontinuity of Boundary conditions 13-10
Discontinuity of Loads 13-9
Discrete Frequencies check box 8-3
Displacement option 2-50
Display button 2-53
Display Options node 2-3
Distributed Load option 2-27
Dynamic Analysis 1-3
Dynamics Setup dialog box 6-4, 8-3

E

Edit tool 2-8
Element properties 13-3
Element Size edit box 2-45
Elements option 2-4
Enforced Motion option 2-32
Event Duration edit box 10-3
Explicit dynamics analysis 12-2
Explicit quasi-static analysis 12-2
Explicit Quasi-Static option 12-5
Extraction Method drop-down 4-4

F

Fatigue Setup dialog box 10-2
Find Thin Bodies tool 2-20
FINITE DIFFERENCE METHOD 13-7
finite element analysis 13-2
Force option 2-24
Frequency Range check box 8-3
Frequency Response Analysis 1-7
From Output option 2-38

G

Gausselimination 13-5
Generate Mesh tool 1-13, 2-47
Geometric Discontinuities 13-8
global stiffness equation 13-5
Gravity option 2-28
Grounded Spring check box 2-16

H

Heat Flux check box 11-2
Heat Flux option 2-37
Heat Generation option 2-36, 11-3
Hooke's Law 1-2

I

Idealizations tool 2-12

Impact Analysis 1-7
Impact Analysis option 7-4
Impact Event Parameters dialog box 7-5
Impact Setup option 7-4
Import from Stress Analysis tool 2-9
Impulse Load 12-3
Initial Condition option 2-32

L

large displacements 3-12
Linear Buckling Analysis 1-5, 5-3
Linear Buckling option 5-3
Linear Equation Based option 2-25
Linear option 2-45
Linear Static Analysis 1-2, 3-2
Linear Steady State Heat Transfer option 11-2
Load dialog box 2-24
Loads tool 1-12, 2-24

M

Manual tool 2-40
Mass Representation drop-down 4-4
Material dialog box 2-9
Material Discontinuity 13-10
material nonlinearity 3-12
Materials tool 1-10, 2-9
Max Element Growth Rate edit box 2-46
Mesh Control tool 2-47
Meshing 2-44
Mesh Settings tool 2-45
Mesh Table button 2-46
Mesh Table dialog box 2-46
Midsurface dialog box 2-21
Midsurfaces tool 2-21
Min Feature Angle edit box 2-46
Modal Database drop-down 4-4
Modal Frequency 8-4
Modal Frequency Response analysis 8-2
Modal Frequency Response Analysis 1-7
Modal Frequency Response option 8-4
Modal Setup dialog box 4-3
Modal Setup option 4-3
Modal Transient Response Analysis 1-6, 6-2
Modal Transient Response option 6-5
Modes Analysis 4-5
mode shape , 4-2
Moment option 2-26
Multi-axial fatigue analysis 10-2
Multi-Axial Fatigue Analysis 1-8
Multi-Axial Fatigue option 10-2

N

natural frequencies 4-2
New tool 2-6
nodal points 13-2
noded isoparametric element 13-3
nodes 13-2
Nodes option 2-3
Nonlinear Buckling Analysis 1-5
Non-Linear Buckling Analysis 5-5
Nonlinear Buckling option 5-5
Nonlinear Material Data dialog box 2-10
Nonlinear Setup dialog box 5-5
Nonlinear Static Analysis 1-3
Nonlinear Static Parameters dialog box 3-17
Nonlinear steady state heat transfer analysis 11-4
Nonlinear Transient Response Analysis 1-6, 6-2
Nonlinear Transient Response option 6-7
Normal Modes analysis 4-2
Normal Modes Analysis 1-3, 4-3
NRL option 9-5
Number of Modes edit box 4-4
Number of Points in the Range edit box 8-3
Number of Timesteps option 6-4

O

Offset Bonded option 2-41
Offset Surface dialog box 2-19
Offset Surfaces tool 2-19
Options tool 2-49
Output Options drop-down 2-8

P

Parabolic option 2-45
Part View check box 2-51
Penetration Type drop-down 2-41
plasticity problems 3-12
Plot dialog box 2-49
Post-Processing option 2-5
PPFA button 2-11
Pressure option 2-28
Prestress Normal 4-5
Prestress Normal Modes option 4-5
Prestress static analysis 3-11
Prestress Static Analysis 1-3
Probes toggle button 2-54
Projectile Translation Vector selection box 7-4
Project Midside Nodes check box 2-46
PSD (Acceleration) vs. Frequency option 9-3
PSD Output Control button 9-2

Q

Quality Midside Adjustment drop-down 2-46
Quasi-Static Setup dialog box 12-6

R

Radiation 1-8
Radiation option 2-36
Random Response analysis 9-2
Random Response Analysis 1-7
Random Response option 9-2
Rayleigh Damping check box 6-4
Refinement Ratio check box 2-46
REFINING MESH 13-11

Remote Force option 2-28
Report Generation option 2-6
resonance , 4-2, 1-4
resonant frequencies 4-2
Response Analysis 8-4
Rigid Body option 2-17
Rod option 2-14
Rotational Force option 2-31
Run button 2-48

S

Save dialog box 2-54
Section View check box 2-50
Select Units button 2-7
Separation/No Sliding option 2-41
Shell Idealization 12-2
Shock/Response Spectrum analysis 9-4
Shock/Response Spectrum Analysis 1-8
Show XY Plot button 2-36
Simulation 1-2
Sliding/No Separation option 2-41
Solver tool 2-43
Specify Min/Max check box 2-50
Spectrum Curve List box 9-6
Spread around Modes check box 8-6
Spring option 2-15
SRSS option 9-5
Static Analysis , 1-2
Steady State Thermal Analysis 1-8
Stiffness check box 2-16
Stiffness Factor edit box 2-42
Strain-Life option 10-3
Stress-Life option 10-3
Stress option 2-50
Structural Members dialog box 2-22
Structural Members tool 2-22
Suppress Short Features edit box 2-46
Surface Contact dialog box 2-40, 2-43
Symmetric Contact option 2-42

T

Table Data dialog box 2-34
Temperature Dependence check box 2-34
Temperature (K) edit box 11-3
Temperature option 2-33
Thermal analysis 1-8
Thermal Convection option 2-34
Time Conversion Factor edit box 10-3
Time Step dialog box 11-5
Time Step(s) option 6-4
Transient response analysis 6-2
Transient Response Analysis 1-6
Transient Thermal Analysis 1-8

U

Units dialog box 2-7
Unsymmetric Contact option 2-41
Upper Jacobian Ratio Bound edit box 2-46
Use Of Symmetry 13-11

V

Variable Load Definition check box 2-25
Vector drop-down 2-50
Vector Options 2-52
Vibrational Fatigue 1-8
Vibration Fatigue Analysis 10-4
Visibility Options tab 2-53

X

XY Plot node 4-7

OTHER BOOKS BY CADCAMCAE WORKS

Autodesk Inventor 2018 Black Book

Autodesk Fusion 360 Black Book (V2.0.6508)

Autodesk CFD 2018 Black Book

ETABS 2016 Black Book

Autodesk Revit 2020 Black Book

AutoCAD Electrical 2020 Black Book
AutoCAD Electrical 2018 Black Book
AutoCAD Electrical 2017 Black Book

SolidWorks 2019 Black Book
SolidWorks 2018 Black Book
SolidWorks 2017 Black Book

SolidWorks Simulation 2019 Black Book
SolidWorks Simulation 2018 Black Book
SolidWorks Simulation 2017 Black Book

SolidWorks Flow Simulation 2019 Black Book
SolidWorks Flow Simulation 2018 Black Book

SolidWorks Electrical 2019 Black Book
SolidWorks Electrical 2018 Black Book
SolidWorks Electrical 2017 Black Book

Mastercam X7 for SolidWorks 2014 Black Book
Mastercam 2017 for SolidWorks Black Book

Creo Parametric 6.0 Black Book
Creo Parametric 5.0 Black Book

Creo Manufacturing 4.0 Black Book

www.ingramcontent.com/pod-product-compliance
Lightning Source LLC
LaVergne TN
LVHW060140070326
832902LV00018B/2882
* 9 7 8 1 9 8 8 7 2 2 7 3 3 *